AF559869

INTRODUCTION TO PHYSICAL GEOGRAPHY

INTRODUCTION TO PHYSICAL GEOGRAPHY

By

Dr. K. Bharatdwaj

DISCOVERY PUBLISHING HOUSE PVT. LTD.

NEW DELHI-110 002

Published by:

DISCOVERY PUBLISHING HOUSE PVT. LTD.
4383/4B, Ansari Road, Darya Ganj
New Delhi-110 002 (India)
Phone : +91-11-23279245; 23253475; 43596065
E-mail : discoverybooksindia@gmail.com
discoverypublishinghouse@gmail.com
namitwasan9@gmail.com
web : www.discoverypublishinggroup.com

First Published: **2009**

Reprinted: **2022**

ISBN: 978-81-8356-440-3

Introduction to Physical Geography

Printed at:
Infinity Imaging Systems
Delhi

Contents

Preface

Study of spatial patterns of weather and climate, soils, vegetation, water and landdorms, including human-environmental interaction is known as physical geography. It covers the topics, relating to the surface of the earth. One of the three major subfields of geography, it focuses on the processes and patterns in natural environment, as opposed to cultural or the built environment, and the domain of human geography.

The four spheres, studied under physical geography include lithosphere, atmosphere, hydrosphere and biosphere. The lithosphere is the rock sphere and comprises the solid and broken rocks, which have been sculpted into landforms at/ near the earth's surface and as well as the thin layer of soil, covering the surface. The hydrosphere is the water sphere and includes water in all its forms (i.e. solid liquid and gas). The atmosphere, the air sphere, a reservoir for heat and moisture, is a complex mixture of gases, including nitrogen, oxygen, argon, ozone, carbondioxide and water vapour. The biosphere is the life sphere and includes all life forms and all areas at earth's surface, capable of sustaining life.

Physical geography can be divided into five main subfields, i.e. climatology, geomorphology, pedology, biogeography and hydrology. Physical geography as a scientific discipline is usually contrasted with and complemented by its sister science, human geography.

Hopefully, this effort of ours would be accorded a warm welcome, among all sections of readership. For further enhancement of the usability of the work, positive and enlightening feedback is solicited.

– Editor

Varied Natural Organism

There is great diversity in the physical environment and living organisms on the earth's surface. The diversity of organisms is a response to the interaction of atmosphere, hydrosphere, and lithosphere, which produce a variety of conditions within which the biosphere exists. This chapter presents a brief account on the systems and cycles of the biosphere — the domain of the living organisms — in which they interact with the hydrosphere, atmosphere and lithosphere. The study of the interactions between organisms (life forms) and their environment is the science of *ecology*.

The sphere of life and organic activity extends from the ocean floor to about 8 km (5 miles) in the atmosphere. The biosphere includes myriad ecosystems from simple to complex, each operating within general spatial boundaries. A group of organisms and the environment with which the organisms interact is known as *ecosystem*. In other words, an ecosystem is a self-regulating association of living plants and animals and their non-living physical environment. In an ecosystem, a change in one component causes changes in others, as systems adjust to new conditions.

Earth's surface environment itself is the largest ecosystem within the natural boundary of the atmosphere. Natural ecosystems are open systems for both energy and matter, with almost all ecosystem boundaries functioning as transition zones rather than as sharp demarcations.

Biogeography deals with the study of the distribution of plants and animals and related ecosystems, the diverse spatial patterns they create, and the physical and biological processes, past and present, that produce this distribution across the earth. Humans are earth's major biotic (living) agent, because human beings influence all ecosystems.

Earth's Ecosystem

The ecosystems of the earth may be divided into: i) aquatic ecosystems, and ii) terrestrial ecosystems.

An aquatic ecosystem is composed of oceans, estuaries, and freshwater bodies. The health of aquatic ecosystem is critical to the biosphere. On the other hand, terrestrial ecosystem is a self-regulating association of plants and animals and their abiotic environment that is characterised by specific plant formations.

Major Terrestrial Biomes of Earth: The major kinds of terrestrial communities are called *biomes* by ecologists. It is a large, stable terrestrial ecosystem characterised by specific plant and animal communities. Each biome is usually named by its dominant vegetation. The main characteristics of each biome differ from the other.

In general, climatic factors exert the primary control over the nature of the *biota* (a collective term for the organisms) of broad geographical areas – they are responsible for the characteristic life forms that permit ecologists to distinguish biomes. The relationship of average annual precipitation, or monthly temperature, and biome is quite complex. Some of the characteristics of ecosystem productivity in the various biomes are listed in table below.

Biological Productivity in Different Biomes

Biomass	*Arctic Tundra*	*Northern Taiga*	*Beech forest (Central Europe)*	*Subtropical deciduous forest (average)*	*Tropical rain-forest (average)*	*Dry savanna (India)*	*Desert*
Photosynthetic parts (%) (mostly leaves)	30	8	1	3	8	11	13

Perennial parts (%) (stems, etc.)	70	73	77	74	47	25	
Roots (%)	70	22	26	20	18	42	62
Total	100	100	100	100	100	100	100
Litter fall (% of plant biomass)	20	4	3	5	8	27	38

Each ecosystem has its own characteristics which individually and collectively influence the biomes. In other words, there are characteristics of the ecosystem in various biomes that transcend the collection of species functioning in the ecosystems. Thus, two tropical rainforests may have entirely different floras, but they will be much more similar to each other in their productivity and in the apportionment of biomass between roots and above-ground parts than either would be to a temperate forest or a desert ecosystem. Geographers recognise 10 to 20 major types of terrestrial ecosystems called *biomes*. A brief account of some of the major biomes is given in the following paras:

Tropical and Equatorial Rainforest Biome: The equatorial and tropical rainforest biome is found in the tropical region of consistent daylight (12 hours), high insolation and average annual temperature around over 25°C (77°F). This biome is characterised with the most diverse body of life on the earth. The Amazon region is the largest tract of equatorial and tropical rainforest, also called the *selva*. In addition, rainforests cover equatorial regions of Congo basin, coastal Gulf of Guinea, margins of Madagascar, South East Asia, Queensland (Australia), the Pacific coasts of Ecuador, and Colombia, and the east coast of Central America with small discontinuous patches elsewhere. The *cloud forests of the western Venezuela* are such tracts of rainforests at high elevation, perpetuated by high humidity and cloud cover. Undisturbed tracts of rainforests are rare.

In the equatorial region, owing to intense heat and high moisture, the growth of trees is dense and luxuriant. Many of the trees grow to tremendous height and in the struggle to reach the sunlight, lianas, for example, climb up other trees. *Epiphytes* are also found abundantly in these forests. Individual species of trees are very scattered but among them are such valuable tropical hardwoods as mahogany, rosewood and ebony. The wild rubber tree is also a native of the equatorial forest.

The smooth, slender trunks of rainforest trees are covered with thin bark and buttresses. These buttresses form angular open enclosures, a ready habitat for various animals. There are usually no branches for at least the lower two-thirds of the tree trunks.

The wood of many equatorial trees is extremely hard, heavy and dense — in fact, some species will not even float (exception is *balsa)*. Logging is difficult because individual species are widely scattered — a species may occur only once or twice per sq km. Selective cutting is required for species-specific logging, whereas pulp production takes everything.

In many places, there are more than 50 species of trees per hectare — many more species than one could find in a day's walk through a temperate forest, where the vast majority of trees represent only a few species. Except where a fallen tree has created a hole in the canopy to admit light, the under-story in a rainforest is sparse, unlike that of a temperate forest. Since the trees normally do not branch below the canopy, one can often walk without almost unobstructed through the dimly lit depths of a mature forest. Most trees have shallow roots, and many develop huge buttresses for support.

Equatorial forests represent approximately one-half of earth's remaining forests, occupying about 7 per cent of the total land area of the world. This biome is stable in its natural state, resulting from the long-term residence of these continental plates near equatorial latitudes and their escape from glaciation.

Animal species in tropical rainforests are also very diverse, with insects, amphibia, reptiles and birds. For example, roughly as many butterfly species can be found in a single rainforest locality as are found in the entire subcontinent of India. Coevolutionary relationships appear to be more highly developed in this biome than in any other.

Tropical forest soils (oxisols) are in general exceedingly thin and nutrient poor. They cannot maintain large reserves of minerals needed for plant growth, such as phosphorus, potassium and calcium, primarily because heavy rainfall and a high rate of water table leach them from the soil. The leaching process leaves behind large residue of insoluble iron and aluminium oxides in the upper levels of tropical forest soils. Soils are essentially infertile, yet they support rich vegetation.

Tropical rainforests are being destroyed more rapidly than any other biome. In Central America, the forests are 'vaporising'. In Brazil, they are being destroyed as a consequence of direct governmental action, in an effort to develop the Amazon basin. Throughout the tropical region, forests are being felled to develop agricultural land. The pressure of human population and ignorance are quickly robbing the world of its richest reservoir of terrestrial organic diversity. The present human assault on earth's rainforests has put this diverse fauna and flora at risk.

Tropical Deciduous Forest and Scrub Biome: On the margins of equatorial biome in the areas of erratic and variable rainfall is the tropical deciduous forest and scrub biome. These areas have three seasons in a year, viz., winter, summer, and rainy seasons. The deciduous broad-leaved trees of this biome drop their leaves before the commencement of scorching summer. Areas of this biome have fewer than 40 rainy days. Heavy monsoon rains are recorded in the ending part of summer (July to September in the Northern Hemisphere).

The monsoonal forest average 15 metres (50 feet) in height with no continuous canopy of leaves, graduating into open orchard-like parkland with grassy openings or into areas choked by dense undergrowth. *Shisham, bargad, mohuwa,* teak, *sal, neem, jamun,* poplar, etc., are the main trees of this biome. In more open tracts, a common tree is *babool* (acacia), with its flat-topped appearance and usually thorny stems. The trees throughout most of this biome are not good for lumber but may be valuable for fine cabinets, especially teak wood. In addition, some of the trees with dry season adaptations produce usable waxes and gums.

Tropical Savanna Biome: Tropical grassland biomes are found between the wet forests and the hot deserts. Many of the ecologists opine that grassland biomes are the zones between forests and deserts where the fire and/or burrowing animals have prevented the spread of trees, even though it is more plentiful than that which produces typical desert.

Savannas covered more than 40 per cent of earth's land surface before human intervention but were especially modified by human-caused fire. Fires occur annually throughout the biome. The timing of these fires is important. Early in the dry season they are beneficial and increase tree cover; if late in the season they are very hot and kill trees and seeds. Savanna trees are adapted to resist the 'cooler' fires. Elephant grasses averaging 5 metres (16 feet) high, and forests once penetrated much farther into the dry regions, for they are known to survive when protected. Savanna grasslands are much richer in humus than the wetter tropics and are better drained, thereby providing a better base for agriculture and grazing. Sorghum, wheat, maize, millets, pulses and groundnuts are the common agricultural commodities.

Africa has the largest region of this biome, including the famous Serengeti plains and the Sahel region. Sections of Australia, Peninsular India, east of Aravalli range (Rajasthan), parts of Brazil, Argentina and Venezuela are all parts of savanna biome.

Some of the local names for these lands include the *Llanos* in Venezuela, the *Campo Cerrado* of Brazil and Guinea, and *Pantanal* of southwestern Brazil.

Large animals typical of grasslands include the bison (buffalo) and pronghorn antelope in North America, wild horses in Eurasia, large kangaroos in Australia, and zebras, giraffes, white rhinoceros, and a vast diversity of antelopes in African savannas. The savannas support the richest fauna of large grazers, and the grazers seem to have coevolved in ways that maximise the utilisation of the local plant resources.

Other important grassland animals include lions, tigers (cheetahs), hyenas, coyotes, and other predators, a variety of birds, ranging in size from ostriches and vultures to small sparrows. Rabbits and other burrowing rodents and grass hoppers (locusts) are also found in savanna grasslands.

The grassland biome has a higher concentration of organic matter in its soil than any other biome. The amount of humus in grassland soil is about a dozen times more than that in the forest soils. The extraordinary richness of grassland soil has led to the establishment of extremely successful agricultural ecosystems in grassland areas – in the American prairies. The agricultural systems can deteriorate, however, if careful soil management is not practised.

The interlaced grass roots and creeping underground stems of grasses form a tuft that prevents the erosion of soil by wind and water. When the turf is broken with a plough, the soil gets exposed to erosive influences and processes. Lack of proper soil conservation has already led to a loss of an estimated one-third of the topsoil of USA and Central Asia. In India, about half of the farmland is not adequately protected from erosion. In fact, in the poorly-fed areas of large population, conservation of soil is difficult.

Agriculture in grassland biomes can only be successful over the long term in areas where considerable effort is put into the maintenance of soil structure and nutrients.

Mid-latitude Broadleaf and Mixed Forest Biome: The mid-latitude broadleaf and mixed forest biome is found in the moist continental climate of Europe, North America and Asia. Relatively lush evergreen broadleaf forests occur along the Gulf of Mexico. Northward are mixed deciduous and evergreen needle leaf forests.

Pines predominate in the southeastern and Atlantic coastal plains. It is also found in New England and westward in a narrow belt to the Great Lakes. White and red pines, hemlock, oak, beech, hickory, maple, elm and chestnut are the main species of this biome.

This biome is well known for its superior timber. The distribution of these species has been greatly altered by human activity. To the north of this biome, poorer soils and colder climates favour stands of coniferous trees and gradual transition to the 'northern needle leaf or boreal forests.

Temperate Rainforest Biome: The temperate rainforest biome is found in the middle and high latitudes along narrow margins of the northwest Pacific (North America). It has few species of trees. It receives about 400 cm (160 inches) precipitation per year on the windward slopes; moderate air temperature, summer fog, and an overall maritime influence produce this moist, lush green vegetation community.

The *tallest trees* in the world occur in this biome – the coastal redwoods. These trees can exceed 1,500 years of age and typically range in height from 60 to 90 metres (200-300 feet), with some exceeding 100 metres (330 feet). Douglas fir, spruce, cedar and hemlock are the other trees of this biome. These species have, however, been reduced to few valleys.

Desert Biome: A desert is an arid region characterised by little or no rainfall in which vegetation is scanty or absent, unless especially adapted or where groundwater conditions are favourable. The term was once confined to hot tropical and subtropical regions where precipitation was greatly exceeded by evaporation, but it is now often used to describe mid-latitude low rainfall areas of continental interiors and also regions of perennial ice and snow of high latitudes where vegetation cannot exist because of low temperatures and physiological drought rather than a deficiency of precipitation. Here, in the desert biome, only the tropical hot deserts have been described and discussed.

The hot deserts are found in the areas where rainfall in less than 25 cm (10 inches). They are concentrated in the vicinity of the latitudes 30° N and 30° S, where the global weather system tends to produce descending masses of dry air. Lack of moisture is the essential factor shaping the desert biome. Most deserts are quite hot in the daytime and, because of the sparse vegetation and resultant rapid radiation of heat, quite cold at night. Some, like the fog desert of the Peruvian coast, which gets little sun over much of the year and the Arctic rock desert are relatively cool even at midday. Both of these desert habitats can be nearly devoid of plants.

Desert plants and animals have evolved many specialisations for conserving water. In plants these include usually thick, waterproof outer layers or cuticle, modifications of breathing pores (stomata), reduction in leaf area, and specialised hairs or outgrowths that reflect light. Many desert plants are also heavily armed with spines to repel the attacks of moisture-seeking animals. They are frequently aromatic, indicating the presence of biochemical defences against herbivores.

Plants may appear to be widely spaced in deserts, but if their roots were visible, the ground between the plants would be seen to be laced with shallow root systems that allow maximum absorption of the rain that does fall. In certain soil conditions, desert plants may have extremely long taproots to reach deep underground water supplies.

Germination of the seeds of desert plants is often inhibited by water-soluble chemicals that must be leached out by a threshold amount of rain before sprouting can occur. Other plants that grow in gravelly *arroyos* have seeds that require abrasion as the first rains wash them along before they will germinate. Thus, deserts appear to turn miraculously green almost overnight following significant rains, and fast-growing annual plants often create spectacular floral displays.

Proportionately more annual plants are found in the desert biome than in any other. If rainfall is relatively abundant, each individual plant may produce numerous flowers. If rain is sparse, many individuals produce only a single blossom each. Since the period of bloom is brief and only follows sporadic rainfall, many people who have travelled in desert areas have not seen them in bloom and remain unaware of the wonderful aesthetic resource they represent.

Desert animals also solve the problem of water shortage in diverse ways. Most of them are active primarily at night, remaining under cover in the heat of the day. Excretory systems are designed to conserve water, and many desert animals are able to use the water they produce in their cellular metabolism. Many desert insects are *'annuals'* like the plants they feed on — synchronising their periods of activity with the evanescent desert boom.

Desert soils contain little organic matter and must ordinarily be supplied with both water and nitrogen if they are to be cultivated. Much of the uncultivated flat land remaining in the world today is desert, and it seems inevitable that large-scale attempts will be made to expand agriculture into this biome. Irrigation in deserts is an expensive and often temporary process — very often a system cannot be maintained and once-cultivated arid lands revert to desert.

Human activities have already produced a great increase in the amount of desert and wasteland. The vast Sahara itself is in part man-made, the result of overgrazing, faulty irrigation, and deforestation, combined with natural climatic changes. Today, the Sahara seems to be advancing southward into the drought-stricken *Sahel,* its advance aided by overpopulation of people and domestic animals. The great Thar desert of the subcontinent of India is also partly the result of human influence. About 3,000 years ago, what is now the centre of the Thar desert was a jungle. The spread of the desert has been aggravated by poor cultivation practises, lumbering, and overgrazing. Human activities could lead to repetition of the Sahara and Thar stories in many other parts of the world.

Mediterranean Biome: Mediterranean biome is confined to coastal areas of the Mediterranean Sea, California, central-southern Chile, southern parts of South Africa, and Australia. This biome is characterised by hot summer and mild-moist winter. Its trees have deep roots which have the ability to obtain moisture from the deep layers.

The dominant shrub formations that occupy these regions are stunted and tough in their ability to withstand hot summer drought. The vegetation is called *sclerophyllous* (from *sclero* or 'hard' and *phyliows* for 'leaf'); it averages a metre or two in height and has deep, well developed roots, leathery leaves, and uneven low branches.

In California, the scrubby evergreen vegetation is known as *chaparral.* Along the Mediterranean Sea this scrub is called as *maquis.* Maquis includes cork-oak trees (source of cork) as well as pine, olive and fig trees. In Chile, such a region is called *mattoral,* in southwest Australia, *malle scrub.* In Australia most of the eucalyptus species is sclerophyllous.

The main crops grown in the Mediterranean biome are citrus fruits, vegetables, nuts (almond, walnut), vines, wheat, oats, barley, artichoker, olive and figs.

Mid-latitude Grassland Biome: Of all the biomes the mid-latitude grasslands are most modified by human activity. It stretches over Steppes (Russia), Prairies of North America, Downs

(Australia), Velds (South Africa) and Pampas of Argentina. These are extensively cultivated and have often been called as the *breadbaskets of the world.* In these regions, the only naturally occurring trees are deciduous broad leafs along streams and other limited sites. These regions are called *grasslands* because of the original predominance of grass-like plants. In each region of the world where these grasslands have occurred, human development of them was critical to territorial expansion.

Cold Desert and Semi-desert Biome: The cold desert and semi-desert biomes tend to occur at higher latitudes where seasonal shifting of the subtropical high pressure belts is of some influence less than six months of the year. Specifically, interior locations are dry because of their distance from moisture sources or their location in rain shadow areas on the lee side of mountain ranges such as the Himalayas, Sierra Nevada, and the Andes. The combination of interior location and rain-shadow positioning produces the cold deserts of the Great Basin of western North America.

Winter snow occurs in cold deserts, but is generally light. Summers are hot, with highs from 30°C to 40°C. Night-time lows, even in the summer, can cool 10°C to 20°C from the day time high temperature. The dryness, clear skies, and sparse vegetation lead to high radiative heat loss and cool evenings. These deserts are the results of overgrazing activity.

Taiga Biome: A Russian term for the belt of coniferous forest which circles the land masses of the Northern Hemisphere between the temperate grassland and the Tundra. Most of the geographers regarded it as synonymous with the *boreal* forests. String bogs is also a characteristic of taiga biome. The conifers have to withstand extremely low temperatures and lengthy periods of snowfall.

These are evergreen forests, as most of the trees do not shed their leaves in winter. Their leaves are specialised to reduce water loss, especially in the cold season when the area is in physiological effect a desert. The leaves in many cases are needle-like and generally last three to five years. In general, the trees are much less diverse than those in temperate forests, and the soils have a different kind of humus and are more acidic.

Boreal forests of pine, spruce, and fir occupy most of the sub-arctic climates on earth that are dominated by trees. Economically, these forests are important for lumbering, with saw timber in the southern margins of the biome and pulp wood throughout the middle and northern portions.

Bears, moose, lynxes, rabbits, squirrels, and a variety of birds are found in taiga biome, but the diversity and abundance of warm-blooded vertebrates are generally less than those in the temperate forests. The diversity of cold-blooded vertebrates is even more dramatically restricted – snakes are uncommon, and few amphibia are to be found. Insect diversity is correspondingly low but the existence of one or two species of conifer leads to outbreak of herbivores like the spruce bud worm, which can defoliate huge areas of forest. Mosquitoes and other biting flies reach abundances in the taiga unknown elsewhere in the world and can make human life very difficult in summer.

Animals' skin for the fur and wood were the major commodities people extracted from the taiga. The demand for fur and wood for pulp and timber have led to an almost total shift to logging in the exploitation of taiga system. Consequently, huge areas have been denuded of forest. Moreover, broadcast use of insecticides against insects has been widespread, adding persistent poisons to the already heavily taxed ecosystems of the earth and normally failing to fulfil the goals of the spray programme.

In the Southern Hemisphere, little of the land area extends to high enough latitudes to support the extensive taiga like that of the Northern Hemisphere. However, there are some cool areas in South America, and Australia, dominated by evergreen trees of the beech fairly in the genus.

Tundra Biome: The treeless plains of northern North America and northern Eurasia lying practically along the Arctic circle and on the northern side of taiga forest is known as the *tundra biome*. There is no corresponding major region in the Southern Hemisphere. For most of the year the mean monthly temperature is below freezing point, and winters are long and severe, the ground being covered with snow. The summers are short and warm, but even in July the mean monthly temperature does not rise above 10°C. The subsoil, i.e., the ground a few centimetres below the surface, may be permanently frozen. This and the strong, intensely cold winds of winter, make normal tree growth impossible, though there may be stunted willows, birches, etc. In summer, mosses and lichens appear in abundance and some flowering plants; much of the flat ground, where drainage is poor, then becomes swampy. *Glei* soils are the characteristics of the tundra.

Caribou, wolves, musk-oxen, arctic foxes, rabbits and polar bears are among the mammals found in the tundra. In the brief Arctic summer, an astounding diversity of birds (especially waterfowl) migrate to the tundra to breed, feeding on an ephemeral bloom of insects and freshwater invertebrates. Mosquitoes can make one's life nearly as miserable in the tundra as in the taiga. Reptiles and amphibians are absent. As in taiga, animal populations in the tundra may be subject to dramatic oscillation in size. Great care must be taken in building on tundra because heat from structures will melt the permafrost and cause uneven settling, which often badly distorts the structures. The tundra has been one of the least exploited biomes, but that era is also ending.

Alpine tundra is similar to arctic tundra, but can occur at lower latitudes, because it is associated with high elevation. This biome is usually described as above the timberline — that elevation above which trees cannot grow. Timberlines increase in elevation equatorward in both the hemispheres. Alpine tundra communities occur in the Andes near the equator, the White Mountains of California, the Rockies, the Alps, the Himalayas, and the Mount Kilimanjaro of equatorial Africa, as well as mountains from the Middle East to Asia. Alpine meadows feature grasses and stunted shrubs, such as willow and heaths.

Altitudinal Gradients of Biomes

The terrestrial communities differ as one goes up into the mountains. This is primarily because the temperature of the air decreases about 6°C for every 1,000 metre increase in altitude, and because, especially in desert areas, rainfall increases with altitude. Thus, in the mountains of Himalayas above 4,000 metres, temperature conditions are similar to those of the tundra biome at sea level and a treeless alpine tundra exist. As one travels in the Greater Himalayas, the tree line gradually lowers while above 5,000 metres there is hardly any vegetation.

Thus, in the mountains it is possible to find compressed in 2,000 vertical metres a series of communities that would occupy a sea level latitudinal gradient 1,000 kilometres or more in length. Alpine habitats contribute greatly to the floral and faunal diversity of the earth because, although in temperature they closely resemble sea level habitats closer to the poles, they differ in other factors, such as day length regime and atmospheric pressure. Anyone who has been active at an altitude of more than 3,500 metres knows that the stresses encountered there are quite different from those at sea level anywhere.

Problems of Deforestation

Despite growing awareness and increasing investment in environmental protection, the shrinkage in forest area and reduction in biodiversity are accelerating. The reduction in forest in the tropical and temperate regions is the cause of great concern for the environmentalists and planners. Deforestation rates in many developing countries, including Brazil, China and India, continue to increase, while in the developed countries many of the forest areas are degraded because of air pollution. Consequently, risk to biodiversity is also increasing.

Although public awareness of the impact of global deforestation has increased in recent years, it has not slowed the rate of deforestation appreciably. A comprehensive assessment of the state of the world's forest, recently released by the Food and Agriculture Organisation (FAO), indicates that the total forest area continues to decline significantly. According to the FAO analysis, deforestation was concentrated in the developing world, which lost nearly 200 million hectares between 1985 and 1995. This loss was partially offset by reforestation efforts, new forest plantation, and gradual regrowth and expansion of forested area (20 million hectares) in developed countries. The result was a net loss of some 180 million hectares between 1985 and 1995, or an average annual loss of 12 million hectares.

Deforestation in Tropics

According to one estimate, about 50 per cent of the earth's existing forests are confined to the tropics. Unfortunately, these forests have been threatened by the over-interaction of man with nature. More than half of the earth's original forest is gone, cleared for pasture and cattle grazing, timber, fuel and farming. On an average, about 1,68,900 sq km forest area is lost each year, and about a third more is disrupted by selective cutting of canopy trees. In these forests, humans set thousands of hectares on fire to clear land for cultivation. Apart from cereals, the shifting cultivators produce cash crops for export, especially rubber, coffee, tea, coconut and sugarcane.

However, poor soil fertility means that the lands are quickly exhausted under farming and are then generally abandoned in favour of newly burned and cleared lands, unless the lands are artificially maintained. Unfortunately, the dominant trees require from 100 to 250 years to establish themselves after major disturbances. Following clearing, the massive growth of low bushes intertwined with vines and ferns means that a new forest may be slow to follow.

Fire related deforestation also rose sharply in Indonesia. In 1997 and 1999, the severe drought helped in the spread of fires, set by plantation workers and farmers into forest areas. In Indonesia, between 1,50,000 to 3,00,000 hectares of forests were destroyed by fire in 1997 to 1999. Although most of the burning took place in secondary forests raiher than virgin rainforest, the impact has nonetheless been high, destroying habitat for a variety of wildlife species from orangutans to tigers. The fires may increase pressure on adjacent virgin forests by increasing access to formerly remote sites.

A study made by the FAO concludes that the leading causes of deforestation are the extension of subsistence farming (more common in Africa and Asia, and government-backed conversion of forests to other land uses such as large-scale ranching, most common in Latin America and also Asia). Poverty, joblessness and inequitable land distribution, which force many landless peasants to invade the forest for lack of other economic means, continue to drive forest clearance for subsistence farming in many regions. Often, people move in forest areas as logging activity creates roads that open formerly inaccessible regions. As for recently planned forest conversion schemes, these are often used to spur short-term economic development, gain better political control of remote forest regions, and expand agricultural output.

Forest's Quality

The state of world's forests is not simply a matter of their extent. Increasing attention is focussed upon the health, genetic diversity and age profile of forests, collectively known as *forest quality*. Logging often does degrade forest quality, inducing soil and nutrient losses and reducing the forest's value as habitat. Logging pressures in many of the remaining large, virgin rainforest areas continue to increase, with logging activity shifting from the largely deforested areas of South East Asia to the rainforests of the Amazon region, Papua New Guinea and the Congo basin.

Forest quality in the developed countries is also of great concern. FAO reports that forest cover in Europe (excluding the former Soviet Union) increased by more than 4 per cent between 1980 and 1995, but forest conditions worsened. Trees are being damaged by fire, drought, pests, and air pollution. More than 25 per cent of trees assessed in a 1995 survey of forest conditions in Europe were suffering significant deforestation. Only about 40 per cent of the European trees are healthy while in India over 65 per cent of the trees are unhealthy.

On the whole, the rapid increase in population growth, rising demand for lumber and fuelwood and the conversion of forests to agricultural land (particularly in Africa) are expected to put increasing pressure on the world's forests in the next few decades. The result will likely be a considerable loss in forest area and quality, with the remaining forest fragmented into smaller isolated tracts.

Different Consequences

Continued forest loss and degradation will have serious socio-economic and ecological implications at the local, regional and global levels. Exploitation and clearance of natural forests are destroying the environment and way of life for tens of thousands of indigenous people. Disappearing of forest cover also represents incalculable losses in biological diversity and ecological services, including nutrient recycling, watershed management, and climatic regulation.

Fragmentation of Forests

Clearing of forests was started by man around 8000 BC when the cultivation of crops was started. Since then about 50 per cent of the forests that covered the earth have been converted to farms, pastures, and other uses. But, the human impact on forests did not stop there. Most of the forests that are left have been heavily altered by humans, often rendered into a patchwork of smaller forested areas. According to a 2001 World Resources Institute (WRI) assessment, just 20 per cent of the earth's original forests remain in large, relatively natural ecosystems – what are known as *frontier forests*.

This fragmentation process is one of the most serious consequences of the current deforestation and degradation of world forests. Frontier forests differ significantly from the dissected, human modified forests that dominate the earth today. Frontier forests are large and natural enough to ensure the long-term survival of their plant and animal species, including the biggest mammals with the most extensive home ranges. As secure habitats for native species, frontier forests are invaluable refuges for global biodiversity. Forests are home to between 50 and 90 per cent of world's terrestrial species – plants and animals that have provided much of the food and other basics that humans need to survive.

Frontier forests take up tremendous amounts of carbon dioxide, and are an important factor in regulating earth's climate. The remaining frontier forests occur either in far northern climates or in the tropics. About 48 per cent of frontier forests are boreal forests (a broad belt of primarily coniferous trees located between Arctic Tundra and the temperate zone), while 44 per cent are tropical forests. By contrast, only a tiny fraction of earth's frontier forests are in the temperate zone.

According to one estimate of the World Resources Institute, 76 countries have lost all of their frontier forests. Another 11 nations are close to losing their last remaining frontier forests, having fewer than 5 per cent of these forests left, all of which are threatened. More than 75 per cent of all frontier forests fall within the three large tracts that cover two blocks of boreal forest: i) Canada to Alaska, and Russia, and ii) one large tropical forest covering Amazon basin, and Guinea. Three countries alone – Brazil, Canada, and Russia – contain nearly 70 per cent of all frontier forests. Logging represents by far the greatest danger to frontier forests.

Frontier Forests Suffer a Variety of Threats

Region	*Percentage of Frontier Forest Under Moderate or High Threat*	*Percentage of Logging*	*Threatened Forest Mining Roads, and Other Infrastructure*	*Frontiers at Agricultural Clearing Removal*	*Risk Excessive Vegetation*	*From Other*
Africa	77	79	12	17	8	41
Asia	60	50	10	20	9	24
North and						
Central America	29	83	27	3	1	14
Central America	87	54	17	23	29	13
North America	26	84	27	2	0	14
South America	54	69	53	32	14	5
Russia and Europe	19	86	51	4	29	18
Europe	100	80	0	0	20	0
Russia	19	86	51	4	29	18
Oceania	76	42	25	15	38	27
World	39	72	38	20	14	13

The majority of frontier forests that are not threatened today lie within boreal regions, inhospitable to most developers. Outside of boreal forests, however, 75 per cent of the world's frontier forests, including all temperate frontier forests, are endangered by human activity.

The felling of trees and deforestation are also adversely affecting the availability of water for drinking, irrigation and industries. Global water consumption rose six-fold between 1900 and 2000 – more than double the rate of population growth – and continues to grow rapidly as agricultural, industrial, and domestic demand increases.

Globally, water supplies are abundant, but they are most unevenly distributed among and within countries. In some areas water withdrawals are so high, relative to supply, that surface water supplies are literally shrinking and groundwater reserves are being depleted faster than they can be replenished by precipitation. Large-scale deforestation in the subtropics, especially in the monsoon areas, and overgrazing in the semi-arid areas, have made them highly vulnerable to droughts and floods. In 2000, in the months of April to June, the greater parts of Gujarat, Rajasthan, Madhya Pradesh, Andhra Pradesh, Orissa, Haryana, and Maharashtra in India, observed a serious drought. The Sahel region of Africa and most of the semi-arid areas of the world remain in the grip of drought after every two years.

According to one study made by the UNO in 1977, about 33 per cent of world's population lives in countries experiencing moderate to high water stress.

Water use in agriculture is slated to increase as world food demand rises. Agriculture already accounts for about 70 per cent of water consumption worldwide, and the United Nations projects a 50 to 100 per cent increase in irrigation water by 2025. Better management of water resources is, therefore, the key to mitigating water scarcity which can be appreciably increased if the afforestation and conservation practises of forests are adopted rigorously.

Various Elements

An ecosystem is a complex of many variables, all functioning independently yet in concert, with complicated flows of energy and matter. An ecosystem includes both biotic (living) and abiotic (non-living) components. Nearly all depend upon an input of solar energy; the few limited ecosystems that exist in dark caves or on the ocean floor depend upon chemical reactions (chemosyn thesis).

Ecosystems are divided into sub-systems, with the biotic portion composed of producers, consumers, and decomposers.

Subdivision and Community

A biotic sub-division within an ecosystem is known as a *community*. A community is formed by interactions among populations of living animals and plants. An ecosystem is the interaction of many communities with the abiotic physical components of its environment. For example, in a forest ecosystem, a specific community may exist on the forest floor, whereas another community functions in the canopy of leaves high above. Similarly, within a lake ecosystem, the plants and animals that flourish in the bottom sediments form one community, whereas those near the surface form another. A community is identified by its physical appearance, the number of species, and the abundance of each, the complex patterns of their interdependence, and the trophic (feeding) structure of the community.

Within a community, two concepts are important: i) habitat, and ii) niche. *Habitat* is a sub-division of the plant environment having a certain combination of slope, drainage, soil type, and other controlling physical factors. It is the specific physical location of an organism, the type of environment in which it resides or biologically adapted to live. In terms of physical and natural factors, most species have specific habitat requirements with definite limits and a specific regimen of sustaining nutrients. *Niche* (French *nicher,* to nest) refers to function, or occupation, of a life form within a given community. It is the way an organism obtains and sustains the physical, chemical and biological factors it needs to survive.

An individual species must satisfy several aspects in its niche. Among these are a habitat niche, a trophic (food) niche, and a reproductive niche. For example, crows are found all over India from Jammu & Kashmir to Kerala and from Rajasthan to Nagaland in human settlements, agricultural lands, forests, pastures and marshlands where they nest. Their trophic niche is grains, seed crops, fruits, insects, flesh, etc., throughout the year. In a stable community, no niche is left unfilled. The competitive exclusion principle states that no two species can occupy the same niche (food or space) successfully in a stable community. Thus, closely related species are separated spatially from one another. In other words, each species operates to reduce competition.

Some species have symbiotic relationships — an arrangement that mutually benefits and sustains each organism. For example, *lichen* (pronounced *'liken')* is made up of algae and fungi living together. The algae is the producer and food source, and the fungus provides structure and physical support. Their mutual beneficial relationship allows the two to occupy a niche in which neither could survive alone.

By contrast, parasitic relationships eventually may kill the host, thus destroying the parasite's own niche and habitat. An example is mistletoe which lives on and may kill various kinds of trees. Some scientists are questioning whether our human society and physical systems of earth constitute a global-scale symbiotic relationship (sustainable) or parasitic one (non-sustainable).

Thus, one may picture the flow of energy through this system as a step by step progression along what is known as a *food chain.*

Ecosystem: Flow of Energy

The flow of energy forms the basis for examining the global productivity of ecosystems, a topic of concern to biogeographers and ecologists alike.

Food Web: Food web is a complex network of interconnected food chains. The organisation of an ecosystem into steps or levels through which energy flows as the organisms at each level consume energy stored in the bodies of organisms of the next lower level.

Food Chains: *"All flesh is grass":* This simple statement summarises a basic principle of biology that is essential to an understanding of both ecological systems and world food problem. The basic source of food for all animal populations is green plants (grass). Human beings and all other animals with which we share this planet obtain the energy and nutrients for growth, development and sustenance by eating plants directly, by eating other animals that have eaten plants (grass), or by eating animals that have eaten animals that have eaten plants, and so forth.

One may think of the plants and animals in an area, together with their physical surroundings, as comprising a system through which energy passes and within which materials move in cycles. Energy enters the ecosystem in the form of radiation from the sun. Through the process of photosynthesis, green plants are able to capture some of the incoming solar energy and use it to bond small molecules into the large (organic) molecules that are characteristic of living organisms. That process is often called *fixing of solar energy.*

The plants and algae in the food web are the primary producers. They use light energy to convert carbon dioxide and water into carbohydrates (long chains of sugar molecules) and eventually into other biochemical molecules needed for the support of life. This process of energy conversion is called *photosynthesis.* Organisms engaged in photosynthesis form the base of the food web.

Net Primary Productivity: The net photosynthesis for an entire plant community is its net primary productivity. This is the amount of stored chemical energy (biomass) that the community generates for the ecosystem. Biomass is the net dry weight of organic material; it is the biomass that feeds the food chain.

The primary producers support the consumers – organisms that ingest other organisms as their food source. At the lowest level of consumers are the *primary consumers* (the snails, insects, and fish). At the next level are the *secondary consumers* (the mammals, birds, and larger fish), which feed on the primary consumers. Still higher levels of feeding occur in the salt marsh ecosystem as marsh-hawks and owls consume the smaller animals below them in the food web. The *decomposers* feed on detritus, or decaying organic matter, derived from all levels. They are largely microscopic [illegible]ms (microorganisms) and bacteria.

Autophyte and Manufacturers

Organisms that are capable of using carbon dioxide as their sole source of carbon are called *autotrophs* or *producers.* They chemically fix carbon through photosynthesis. Organisms that depend on autotrophs for their carbon are called *heterotrophs* or *consumers.* Plants are the essential producers in an ecosystem – capturing light and generating heat energy and converting it to chemical energy, incorporating carbon, forming new plant tissues and biomass, and freeing oxygen, all as a part of photosynthesis.

From these producers, which manufacture their own food, energy flows through the system along a circuit called *food chain,* reaching consumers and eventually decomposers. Ecosystems are generally structured in a *food web,* a complex network of interconnected food chains. In a food web, consumers participate in several different food chains. Organisms that share the same basic foods are said to be at the same trophic (feeding nutrition) level.

The Users

Primary Consumers: The primary consumers feed on producers (plants). The primary consumers are also called as *herbivores* or *plant eaters.*

Secondary Consumers: The secondary consumers eat animals' meat and therefore they are called as *carnivores.*

Tertiary Consumers: A tertiary consumer eats primary, and secondary consumers and is referred to as the *top carnivore* in the food chain, like the sperm whale.

Omnivore: A consumer that feeds on both producers (plants) and consumers (meat) is called *omnivore* – a role occupied by humans, among others.

Decomposers: The microorganisms that digest and recycle organic debris and waste in the environment include bacteria, fungi, insects and worms.

Abiotic Elements

The abiotic (non-living) components in each ecosystem are very important as they help in the flow of energy and the cycling of nutrients and water. They set the stage for ecosystem operations.

Light, temperature, water and climate vary from latitude to latitude, and they vary altitudinally also. These variations influence the photosynthesis, photoperiod, and chemical reactions in each of the ecosystems. In other words, there is a direct relationship between temperature, precipitation and vegetation.

There is vertical zonation also in vegetation. This fact was appreciably recognised by Alexander von Humboldt in the early part of the 19th century. While exploring the Andes mountains of Peru, he deduced that plants and animals recur in related grouping wherever similar conditions occur in the abiotic environment. He described a distinct relationship between altitude and plant communities and developed the *life zone* concept. As he climbed the mountains, he noticed that the experience was similar to that of travelling away from the equator towards higher latitudes.

Beyond these general conditions, each ecosystem further produces its own micro-climate, specific to individual sites. For example, in forests the insolation reaching the ground is reduced. A pine forest cuts light by 20 to 40 per cent, whereas an equatorial forest reduces it by 60 to 80 per cent. Forests are also 5 per cent more humid than non-forested landscapes, have warmer winters and cooler summers, and experience reduced winds. Such highly localised micro-ecosystems are evident along a mountain range, where changes in exposure and moisture can be easily seen.

The net primary production and plant biomass in the various regions of land, oceans, lakes, swamp, marsh, and streams have been given in table below.

Net Primary Production and Plant Biomass on Earth

Ecosystem	*Area (10^6 Sq km)*	*Net Primary per Unit Area Normal Range*	*Productivity (g/sq m/yr) * (mean)*	*World Net Primary Production (10^2/t/yr)***
Tropical rainforest	17.0	1000-3500	2200	37.4
Temperate deciduous forest	7.0	600-2500	1200	8.4
Boreal forest	12.0	400-2000	800	9.6
Woodland and shrubland	8.5	250-1200	700	6.0
Savanna	15.0	200-2000	900	13.5
Temperate grassland	9.0	200-1500	600	5.4
Tundra and alpine region	8.0	10-400	140	1.1

Desert and semi-desert scrub	18.0	10-250	90	1.6
Cultivated land	14.0	100-3500	650	9.1
Swamp and marsh	2.0	800-3500	2000	4.0
Lake and stream	2.0	100-1500	250	0.5
Open ocean	332.0	2-400	125	41.5
Algal beds and reefs	0.6	500-4000	2500	1.6
Estuaries	1.4	200-3500	1500	2.1

* 1 sq km - 0.39 sq mi; ** 1 g per sq m - 8.9 Ib per acre; *** 1 metric ton - 1.1023 tons.

Nutrient Components

The most abundant natural elements in living matter are hydrogen (H), oxygen (O), and carbon (C). These elements collectively make up more than 99 per cent of earth's biomass. In fact, all life (organic molecules) contains hydrogen and carbon. In addition, nitrogen (N), calcium (Ca), potassium (K), magnesium (Mg), sulfur (S), and phosphorus (P) are the significant nutrient elements necessary for the growth and development of a living organism.

Ecology Relationship

There is a close relationship between the biotic and abiotic elements of any ecosystem. The energy flow in them has been described in the preceding paras. According to the first law of thermodynamics, *energy can neither be created nor destroyed,* although it may be changed from one form to another (as in the change from light energy to the energy of chemical bonds in photosynthesis). The second law of thermodynamics says, in essence, that *in any transfer of energy there is a loss of available energy,* that is, a certain amount of the energy is degraded from an available, concentrated form to an unavailable, dispersed form. The practical expression of this law as it applies to food production is that no transfer of energy in a biological system is 100 per cent efficient, some energy always becomes unavailable at each transfer. In the photosynthesis system, usually 1 per cent or less of the sunlight falling on green plants is actually converted to the kind of chemical bond energy that is available to animals eating the plants. Usually, about 10 per cent of the energy in the plants is converted into chemical bonds. An equivalent amount of energy may, in turn, be incorporated into chemical bonds of other animals that eat the animals that ate the plants.

At each transfer of energy in a food chain, about 90 per cent of the chemical energy stored in the organisms of the lower level becomes unavailable to those of the higher level (in certain cases, the percentage lost may be much greater or less than this). Since the total amount of energy entering the food chain is fixed by the photosynthetic activity of the plants, obviously more energy is available to organisms occupying lower positions. For example, as an oversimplification, it might take roughly 10,000 kg of grain to produce 1,000 kg of cattle, which in turn could be used to produce 100 kg of human beings. By moving people one step down the food chain, ten times as much energy would be made directly available to them — that is, the 10,000 kg of grain used to produce 1,000 kg of cattle could be used instead to produce 100 kg of human beings.

Succession of Ecology

It is a process whereby different and usually more complex assemblage of plants and animals replace older and usually simpler communities. Changes apparently move towards a more stable and mature condition. In other words, ecological succession (sequence) of distinctive plant and animal communities occurs within a given area of newly formed land or land cleared of plant cover by burning, clear cutting, or other agents. Besides, it takes place when older communities of plants and animals (usually simpler) are replaced by newer communities (usually more complex).

Biological communities are not static as they change in space and time. The change, however, occurs over relatively long periods of time — hundreds, thousands, and millions of years — as the climate of earth changes and their component species evolve and coevolve. They also change on a time scale of years, decades, and centuries, in a process known as *succession.* Suppose, a piece of land is newly exposed by the threat of a glacier, the lowering of a lake, or the emergence of land from the ocean by volcanic action. Gradually, the exposed rock is fragmented under the action of the wind, running water and alternate heating and cooling, and soil starts to form. Plants such as lichens invade and speed the process of soil formation through chemical interaction. Grasses and herbs then move in, adding the action of their roots (and their decaying remains) to the process. Herbivores arrive, followed by carnivores that feed upon them; then the process of primary succession is well under way.

Primary succession is the development of a biological community where none existed previously. The terrestrial process begins with bare substrate and, if undisturbed, continues until a relatively stable community characteristic of the climate regime and the soil type develops. Such a community is called *climax.* The entire sequence of communities from bare ground to the climax is called a *sere,* and the intermediate communities are called *serel stages.*

Secondary succession occurs when a climax community is destroyed, say, by a fire or human activities. It differs from primary succession largely in the early stages. Human beings are often responsible for secondary succession, as a clearing in a tropical forest is abandoned by shifting cultivators (*jhuming, milpa,* etc.).

Successional stages may be interrupted by fire. It is estimated that one-fourth of the earth's land area experiences fire each year. The Sumatra fire in 1999 created numerous ecological and environmental problems.

Lakes and ponds exhibit another form of ecological succession. A lake experiences successful stages as it fills with nutrients and sediment and as aquatic plants take root and grow, capturing more sediment and adding organic debris to the system. The gradual enrichment of water bodies is known as *eutrophication.*

The Backdrop

The earth is more than 4,500-million-years old. Its origin is shrouded in mystery and almost nothing is known of conditions during the first 1,000 million years before consolidation of its crust. The historians, relying on archaeological discoveries and the written records of ancient civilizations can reach back a mere 6,000 years into the past, but the geologists can reconstruct the story in considerable detail for almost 600 million years before the historians' record begins. The geologists' evidence lies in the rocks and in the fossilised relics of plants and animals that many of the rocks contain.

Discovery of Time

Geologic time was discovered in Edinburgh (Scotland) by a small group of scholars led by James Hutton (1726-1797). These scholars challenged the conventional thinking of their days in which the largest unit of time was the human life span (the lives of the patriarchs), and the age of earth was accepted to be 6,000 years, as established by Bishop Usher's summation of biblical chronology. Hutton and his friends studied the rocks along the Scottish coast and observed that every formation, no matter how old, was the product of erosion from even older rocks. Their discovery showed that the roots of time were far deeper than anyone had supposed. Hutton's discovery of time was based on the interpretation of rocks as products of events in earth's history. It was perhaps the most significant discovery of the 18th century because it changed forever the way we look at earth, the planets, the stars, and, as a consequence, the way we look at ourselves.

Theory of Uniformitarianism

The principle of uniformitarianism, initially proposed by James Hutton (1726-97) in his *Theory of the Earth* (1795) and later amplified by Charles Lyell in *Principles of Geology* (1830), states that the laws of nature do not change with time. We assume that the chemical and the physical laws in force today have operated all through the time. Oxygen and hydrogen, which today combine under certain conditions to form water, did so in the past under those conditions. Although scientific explanations have improved and changed over the centuries, natural laws and processes are constant and do not change. All chemical and physical actions and reactions occurring at present are produced by the same cause that produced similar events 100 years or 5 million years ago. *"The present is the key to the past"* describes the principle of uniformitarianism, and therefore, it has been considered as its definition.

In the opinion of Hutton, the earth had evolved by uniform, gradual processes over an immense span of time, and he developed a concept that became known as the *principle of uniformitarianism.* According to the principle, past geologic events can be explained by natural processes we observe operating today, such as erosion by running water, volcanism, and the gradual uplift of earth's crust. Hutton assumed that these processes occurred in the distant past just as they occur now. He saw that in the vast abyss of time, enormous work could be achieved by what appeared to be small and insignificant processes. Rivers could completely erase a mountain range. Volcanism and earth movements could form new ones. On the basis of his observations of the rocks of Great Britain, he visualised "no vestige of a beginning — no prospect of an end". In a way, what Copernicus did for space, Hutton did for time. The universe does not revolve round the earth, and time is not measured by the life span of the humans. Before Hutton, human history was all of history. Since Hutton, we know that we are but a tiny pinpoint of an extraordinary long time line.

Sir Charles Lyell (1797-1875) based his *Principles of Geology* on Hutton's uniformitarianism. Lyell's book established uniformitarianism as the accepted method of interpreting the geologic and natural history of earth. Charles Darwin (1809-82) also accepted Lyell's principle in formulating his theory of the origin of species and the descent of man. Lyell, however, believed that the rates at which processes operate do not change with time. Moreover, recent studies indicate that changes in the rates of various processes may have occurred within certain limits.

With the help of modern scientific instruments, geologists have studied much more of the geologic record than did Hutton and Lyell, and they have observed and measured many subtle details that earlier scientists could not measure. Modern science is making significant advances in understanding earth, its long history, and how it was formed. By applying principles of thermodynamics, electromagnetism, chemistry and related scientific disciplines, geologists are discovering more about earth's genesis and evolution. As a result, the principle of uniformitarianism has been verified innumerable times.

Many features of rocks serve as records or documents of past events in earth's history; for those who listen, the rocks still echo the past. Igneous rocks are records of thermal events: the texture and composition of an igneous rock indicate whether volcanic eruption occurred or if the magma cooled beneath the surface. Sedimentary rocks record changing environments on the earth's surface — the rise and fall of sea level, changes in climate, and changes in life forms. A layer of coal is a record of lush vegetation growth, commonly in a swamp. Limestone, composed of fossil-shell debris indicates deposition in shallow sea. Salt is precipitated from sea water or from saline lakes only in an arid climate so a layer of salt carries specific connotations. The list of examples could go on and on. For more than two centuries, geologists have extracted from the rocks a remarkably consistent record of events in earth's history — a record of time.

Geologic time is, however, continuous; it has no gaps. In any sequence of rocks, however, there are many major discontinuities (unconformities) that indicate significant interruptions in the rock-forming processes.

Theory of Catastrophism

The hypothesis, now more or less completely discarded, that change in the earth occurs as a result of isolated giant catastrophes of relatively short duration, as opposed to the idea implicit in uniformitarianism that small changes are taking place continuously.

Catastrophism holds that earth is young and that mountains, canyons and plains were formed through catastrophic events that did not require aeons of time. Because there is little physical evidence to support this idea, Catastrophism is more appropriately considered a belief than a serious scientific hypothesis. Ancient landscapes, such as the Appalachians and Aravallis, represent a much broad time scale of events.

However, geologic time is punctuated by dramatic events, such as massive landslides, earthquakes, and volcanic episodes. Within the principle of uniformitarianism, these localised catastrophic events occur as small interruptions in the generally uniform processes that shape the slowly evolving landscape.

The Standard Geological Time Scale: Using the principles of superposition and faunal succession, geologists have determined the chronological sequences of rocks throughout broad regions of every continent and have constructed a standard geological time scale that serves as a calendar for the history of the earth. The standard geologic column was established from studies of rock sequences in Europe which is now used worldwide. Rocks are correlated from different parts of the world on the basis of the fossils they contain.

The nomenclatures of the columns indicate eras, periods and epoch in earth's history. These names mark highlights in the historical development of geological science over the past 200 years. Nearly every name is associated with an important scientific discovery. An understanding of the origin and meaning of these names is therefore helpful. It is worthwhile to mention that absolute time designates a specific duration of time in units of hours, days, or years. In geology, long periods of absolute time can be measured by radio metric dating.

The geological time hierarchy comprises of: aeon, era, period, epoch, age, and chron (in descending order of magnitude). The lithostratigraphical (stratomeric) hierarchy is based on: system (period), series (epoch), stage (age), and chrono zone (chron). There are, however, no stratomeric equivalents to aeon and era. The individual time intervals are not of standard duration nor are the rock divisions of uniform magnitude.

Chronostratigraphy and Lithostratigraphy

Chronostratigraphy	*Lithostratigraphy (stratomeric hierarchy)*
Aeon (10 years)	–
Era	–
Period	System
Epoch	Series
Age	Stage
Chron	Chrono zone

The geological time for the purpose of determining the age of various rocks has been divided into two aeons, viz., (i) Precambrian and (ii) Phanerozoic, and four major eras, viz., (i) Precambrian, (ii) Palaeozoic (ancient life), (iii) Mesozoic (middle life), and (iv) Cainozoic or Cenozoic (recent life). The oldest and by far the longest is the Precambrian which covers about 87.6 per cent of the geological time. Our knowledge about this era is quite meagre but our knowledge about the remaining three eras, starting about 600 million years ago, is relatively more.

Each of the eras is sub-divided into smaller time units known as *periods* (systems). The Palaeozoic has seven periods, the Mesozoic three periods and the Cainozoic two periods. Each of the twelve periods is further sub-divided into epochs, each epoch into ages, and each age is divided into chrons. The usage of chron is, however, now obsolete.

Precambrian Era: This is the oldest era of geological history. It is the period of time from the consolidation of the earth's crust to the base of Cambrian. The terms Proterozoic, Azoic and Archaen have been used either as synonyms or partial synonyms.

The duration of Precambrian time is probably not less than 4,000 million years or over 90 per cent of the geological time and during this period a number of orogenies are known to have occurred. The Precambrian rocks are mostly crystalline igneous and metamorphic which are extremely complex. The peninsular plateau of India is mostly composed of the Precambrian rocks.

The hot climate of the Precambrian period is broken up by a series of ice ages. Sea weeds are the only form of vegetation. During the Precambrian period life originated in the warm seas (how it originated is still a mystery). There was no life on land. Thus, the Precambrian rocks contain only a very few fossils of the more primitive forms of life.

The Precambrian rocks are extremely rich in metallic minerals, and almost all the important occurrences of iron ore, gold, silver, copper, manganese, uranium, chromium, lead, zinc and mica in the world are found in the rocks of Precambrian era. These rocks are, however, poor in the fossil fuels like, coal, petroleum and natural gas.

Palaeozoic Era (Ancient Life): The era ranged in time from 600 to 230 million years – a duration of 370 million years. It comprised the Cambrian, Ordovician and Silurian systems in the older or lower Palaeozoic sub-era, and the Devonian, Carboniferous and Permian systems in the newer or upper Palaeozoic sub-era. The boundary between the lower and upper Palaeozoic is drawn at 400 million years. The Palaeozoic was preceded by the Precambrian era and followed by the Mesozoic era. In older works, it may sometimes be referred to as the 'Primary Era'.

The major orogenies that occurred during the Palaeozoic era are: the Caledonian in the lower Paleaeozoic and the Variscan at the top of the upper Palaeozoic, together with their accompanying granite intrusion and the metamorphism.

Cambrian Period: The word 'Cambrian' came from Cambria, the Latin name of Wales, where these rocks were first studied. In most areas of the world, Cambrian rocks rest on the highly deformed Precambrian metaporphic complex. This period started about 570 million years ago and had a duration of about 70 million years. This period was characterised by shallow seas which covered much of the earth. The seas tended to advance on and retreat from the land areas.

Cambrian rocks are found in Wales, northwest Scotland, and western England. Similar rock formations also occurred in Canada and USA (Grand Canyon region). There had been considerable volcanic activity in Europe, but there is no evidence of important mountain building in this period. During the Cambrian period climatic conditions were moderately warm and equable, plant life was confined to sea and all the major groups of invertebrates evolved. There was no life on land.

Ordovician Period: Ordovician had been named after the Ordovices – an ancient Celtic tribe of Central Wales. The period extended from 500 to 435 million years – a duration of 65 million years.

This period was characterised by widespread vulcanicity and the onset of Caledonian orogeny is evident. The seas continued to advance and retreat.

Rocks of Ordovician period have been found not only in Wales but also in parts of North America and northwest Europe. Volcanic eruptions occurred on the floor of the sea. The climate appeared warm and, with no marked climatic zone. Plant life was mainly confined to sea. All life was still restricted to water, but first vertebrates appeared. Plant life did not advance beyond sea weeds. No life found on land.

Silurian Period: The Silurian period had been named after the Silures – also an ancient Celtic tribe of the Wales. Rocks of this period are exposed on the border of Wales. The period extended from 435 to 395 million years, having a duration of 40 million years.

During this period, the level of the seas tended to rise and fall periodically, causing regular changes in the land areas. Rocks of this period are found in Shropshire (UK), Baltic region (Europe) and Niagara Falls (Canada and U.S.A). There was less volcanic activity in this period than that in the

Ordovician Period: Generally warm and equable climate persisted, but was exceptionally dry in certain areas.

Plants first adapted themselves to life on land, but were still leafless. Fossil remains of land vegetation have been found in Australia. New species of vertebrates developed in the sea. Coral reefs developed on a large scale. First plants appeared on land.

Devonian Period: This period has been named after the country of Devon in southwest England. It extended from 395 to 345 million years, having a duration of 50 million years.

During this period, land area increased at the expense of sea. This was a period of extensive mountain building and volcanic activity. Old red sandstone of Europe developed during this period. It was warm and semi-arid climate in northwest Europe and over a large part of North America, with heavy seasonal rains. Equable conditions prevailed elsewhere.

Earth began to look green as plants with roots, stems, and leaves evolved. There was rapid evolution of vertebrate animals. Ancestors of all modern fish evolved. Primitive sharks, measuring upto 7 metres (21 feet), appeared. In consequence, this period has become known as the *Age of Fish*. By the end of the period, the first amphibian animals had come into existence.

With land plants to feed on, the first invertebrate animals left the sea and adapted themselves to life on land. They included millipedes, miter, spiders, and wingless insects.

Carboniferous Period: This is a period named (by Conybeare, 1822) from the widespread occurrence of carbon in the form of coal in these beds. That is why, it is sometimes referred to as the 'Coal Age'. It extends from 345 to 280 million years, and has a duration of 65 million years. Carboniferous rocks are subdivided into two major units: The Pennsylvanian and the Mississippian (named after the Upper Mississippi valley).

Clear, shallow seas were widespread in early period, most of Europe and large parts of Russia were under water. Later sea beds began to rise, exposing great stretches of land. Other land areas sinked, producing brackish swamps over much of Europe and North America. This was the main coal-forming period, particularly in the Northern Hemisphere.

The land climate was extremely dry throughout most of the period, but in some regions it was warm and moist enough to encourage dense vegetation.

Giant evergreen trees, reaching heights of over 33 metres (100 feet), flourished in the tropical swamp of the period. Amphibian creatures continued to develop. Marine life, both plant and animal, abounded in many varieties. The reptile became the first creature to breed on land. Certain species of insects developed wings.

Permian Period: The Permian period was named at the suggestion of Murchisan in 1841 from the province of Perm in Russia. This period extended from 280 to 225 million years — a duration of 55 million years. It marked the end of the Palaeozoic era. This was a period of considerable earth movements. Lofty mountains formed in Europe, Asia and eastern USA (Applachian).

Contrasting climate conditions persisted, mainly arid in the Northern Hemisphere, with occasional warm and humid zones, but ice age conditions prevailed over much of the Southern Hemisphere.

As seasonal variations in climate and temperature developed, evergreen plants began to decrease in number, and deciduous plants, able to withstand periods of drought and forest appeared.

This period marked the end of dominance by marine creatures, as animal and plant life on land increased. Creatures capable of living on land increased in number and variety. A great variety of insects began to emerge.

Mesozoic Era (Middle Life): A major division of geological time extending from c. 250 to 65 million years ago, can be subdivided into three periods: i) Triassic, ii) Jurassic, and iii) Cretaceous. The Mesozoic era was preceded by the Palaeozoic and followed by the Cainozoic (Tertiary) era. The most spectacular elements of the Mesozoic faunas were the giant reptiles. The first mammals appeared at the commencement of the era.

Triassic Period: The Triassic period has been named after the three-fold mountain in Germany. The period extended from 225 to 195 million years, having a duration of 32 million years. It marked the beginning of the Mesozoic era.

Deserts and shrubs-covered mountains made up most of the earth's land area. Formation of marl and sandstone deposits in the warm seas. Hot dry conditions prevailed almost everywhere. Climate became wetter towards the end of the period.

The first carnivores, fish-shaped reptiles, evolved in this period. So did flying fish and the first lobster-like creatures. The reptiles continued to dominate life on land. Dinosaurs, no more than six inches long, appeared for the first time. The first flies and termites also came on.

Jurassic Period: The Jurassic period has been named after the Jura mountains of France by Von Humboldt in 1795. It extended from 195 to 135 million years, having a duration of 60 million years.

The seas advanced again. Most land areas consisted of forests or swampy plains with lakes and meandering rivers. Much of Asia and Europe, including the neighbourhood of Britain, was invaded by the sea.

The climate was predominantly mild, becoming sub-tropical in some regions later in the period. There was sufficient rainfall to support luxuriant vegetation. Conifers, ferns and tree-ferns continued to flourish. Some cycads had flower-like cones — the first step in the evolution of flowers.

Reptiles increased in size and variety. The first bird evolved feathers from scales but retained teeth, solid bones, and jointed tails. Mammals remained small and primitive, no bigger than rat and lived mostly in woodland.

Cretaceous Period: This period has been named after Greek word, *creta* meaning 'chalk'. The Cretaceous period had a duration of about 72 million years, from 136 to 64 million years.

Land areas bordering the sea consisted of far-reaching swamps. The rivers started flowing slowly and formed enormous deltas. There was widespread deposits of chalk (Dover and Dorset of UK, etc.). The climate continued to be mild, causing vegetation to grow abundantly as far north as Greenland, though parts of Australia appeared to be covered by glaciers. Growth of deciduous trees like fig, magnolia, poplar, and plane. Giant turtles and sea-serpents developed.

Giant reptiles, dinosaurs dominated life on land. By the end of the period dinosaurs became extinct.

Cainozoic Era (Modern Life): The division of geological time which succeeds Mesozoic era and ends at the Quarternary. The duration of Cainozoic era was approximately 63 million years, from 65 to 2 million years. It is commonly used as a synonym for Tertiary.

Palaeocene and Eocene Epochs: Palaeocene is the lowermost division of the Tertiary system, while Eocene is the epoch of the Tertiary period between the Palaeocene and Oligocene epochs. Both started about 60 million years ago and lasted for about 27 million years.

Mountain ranges which began to form in Cretaceous period continued to grow. Volcanic activity led to the formation of the Atlantic and Indian oceans. Vast amount of lava was deposited in the Arctic, Scotland, Ireland and Deccan plateau of India.

Flowering plants, including deciduous trees, became dominant. The warm climatic zone right upto Greenland, allowed palms to grow in the region and Malaysia-type jungle in the region of London.

Most species of fish in the ocean took on the shape we know today. Many varieties of modern mammals came into existence, i.e., ancestor of elephants, the rhinoceros, the horse, the pig and the cattle. Primitive monkeys and gibbons appeared in Myanmar (Burma).

Oligocene Epoch: Oligocene epoch started about 38 million years ago and lasted for about 12 million years. Throughout this period the land mass grew at the expense of the sea. The Alps began to form. The forests dwindled and grassland spread, leading to an increase in grass-eating mammals.

The ancestors of modern cats, dogs, and bears evolved. A tailless primitive ape, possibly related to the ancestors of man also appeared.

Miocene Epoch: Miocene epoch started about 26 million years ago and lasted for about 19 million years. During this period powerful earth movements led to a further retreat of the sea. The

Mediterranean became virtually a landlocked ocean. The European and Asian land masses were finally joined together. Increased rainfall led to intense erosion in some parts. Further powerful movements in the earth's crust completed formation of the Alps and led to formation of the Himalayas. There was also much volcanic activity.

Sharks, particularly abundant during this period, grew to enormous sizes, measuring over 20 metres in length and having teeth 6-15 cm long. Increase in the number of apes. Elephants steadily increased in size. Primitive penguins, some as tall as man, lived in Antarctica.

Pliocene Epoch: Pliocene epoch started about 7 million years ago and lasted for about 5 million years. Continents and oceans began to take on their present form. Land subsidence contributed to formation of the North Sea, the Black Sea, the Caspian Sea, and the Sea of Aral.

Giant sharks became extinct, Man-like apes, *Australopithecus* which walks upright spread. Elephants also thrived.

Pleistocene Epoch: Pleistocene epoch started about 2 million years ago and lasted for about 2 million years. Ice sheets and glaciers covered most of the Europe, America, Antarctica and the Himalayas. Melting ice formed the Great Lakes of North America. The tremendous weight of retreating glaciers cut out the fiords of Norway. This was the period of abnormal and extreme climatic change. Marine life was much as it is today.

Ape-like creatures developed enough intelligence to make stone implements for cutting up animals they had killed. Primitive man spread to Asia and Europe. True elephants, horses, oxen first appeared.

Holocene Epoch (Neolithic Period): Holocene epoch started about 10 thousand years ago. The ice continued to retreat, causing the sea level to rise further. Climatic conditions gradually became more equable. In North Africa and Middle East, increasing dryness produced deserts.

Marine life was much as today. Man learned to domesticate animals and cultivate plants.

Magnitude of the Age of the Earth: As stated at the outset, the age of the earth is about 4,500 million years. Man appeared on the earth surface very late in the history of the earth. It has been calculated that if one thinks of the entire history of the earth as having occurred in one year, the first known living creature would have appeared about 240 days before the end of the very last day, and modern man some five minutes before midnight of the last day of the year.

Meaning of Geochronology

The determination of the age of earth is known as *geochronology*. The age of the earth is determined with the help of (i) salinity of the ocean, (ii) sedimentation, (iii) rate of erosion, and (iv) radioactive elements.

Salinity of the Ocean: The water of the ocean is saline. The salinity varies from about 2.5 per cent in the Arctic Ocean to 3.9 per cent in the Red Sea. It is believed that the oceans at the time of their creation would have contained pure water or there was no salt in the sea water. Later on the running water removed salt contents from the land areas and carried that to the seas and oceans. If the total amount of oceanic salt is known and if the annual rate of increase of oceanic salinity is ascertained, the age of the oceans may be determined. On the basis of this technique John Joly (1899) concluded that the earth was 90-100 million-years old. The oceans were created about 80 million years ago. Joly's estimates were far too low because he failed to take into account the fact that salt is removed from the sea and deposited as part of marine sediment. Later, the salt may be exposed by uplift, eroded, and recycled back to the sea.

Sedimentation: The sedimentary rocks are deposited by the various agents of denudation. The process of formation of sediments and sedimentary rocks continued throughout the geological periods and the processes of erosion and deposition are still continuing. The continuous sedimentation resulted into thickening of sedimentary rocks. If one could find out the total thickness and the annual rate of deposition of sedimentary rocks, then the age of the formation of the first sedimentary rocks on the earth surface may be determined which may help in the estimation of the age of the earth. This method, however, does not give reliable and satisfactory results. There are two objections about this method of determining age of different rocks. First, accurate estimates of the average rate of sedimentation are difficult to obtain because different kinds of sediments accumulate at vastly different rates. Second, many interruptions occur in the sequence of sedimentary rocks. In a given area, sedimentation will be followed by uplift and erosion, subsidence, and then renewed sedimentation. An unknown amount of time is not recorded by uplift and erosion, and during those processes, part of the previously formed record is removed.

Rate of Erosion: Some of the geologists opine that the age of the earth may be calculated on the basis of the rate of erosion of the land masses. If we can find out the total amount of denudation of the surface materials till now and the annual rate of denudation, then the age of the earth may be estimated. This method is, however, not very reliable in determining age of the earth.

Radioactive Elements: Lord William Kelvin reasoned that the earth cooled from a molten state and that the entire rate of cooling could be determined by measuring the present rate of heat flow. The earliest methods, utilising uranium and thorium as the starting minerals, yielded evidence that the extent of the geological time was at least 2,000 million years. The oldest dated rocks (3,900-million-years old) are found in west Greenland. By this method it is effectively certain that rocks exist with an age of greater than 4,000 million years. Another method based on the relative abundance of various isotopes of lead in galena (Pbs) yield figures of 5,000 to 5,400 million years.

Astronomical data suggest an age for the solar system of 5,000 ± 1,000 million years, which is in reasonable agreement with geochemical evidence. Rock samples from the moon have been dated at 4,700 million years, also in good agreement with these estimates.

In brief, these and other early attempts to measure geologic time were remarkable first efforts. Each method showed that the earth was far older than anyone had supposed, but the true dimensions of time remained elusive.

Comparative Dating

Relative dating determines the chronological order of a sequence of events. The most important methods of relative dating are based on:

1. The Principle of Superposition.
2. The Principle of Faunal Succession.
3. The Principle of Cross-cutting Relation.
4. The Principle of Inclusion.

The Principle of Superposition: The principle of superposition is the most basic guide in the relative dating of rock bodies. It states that in a sequence of under formed sedimentary rock, the oldest beds are on the bottom and the higher layers are successively younger. The relative ages of rocks in a sequence of sedimentary beds can thus be determined from the order in which they were deposited.

In applying the principle of superposition, we make two assumptions: 1) layers were essentially horizontal when they were deposited, and 2) the rocks have not been so severely deformed that the beds are overturned (Rock sequences that have been overturned are generally easy to recognise by their sedimentary structures, such as cross-bedding, ripple-marks, and the mud-cracks).

The Principle of Faunal Succession: The principle of faunal succession states that groups of fossils of animals and plants occur in the geologic record in a definite and determinable order and that a period of geologic time can be recognised by its characteristic fossils. Thus, in addition to superposition, the sequence of sedimentary rocks in earth's crust is characterised by another independent element that can be used to establish the chronological order of events.

Fossils are the actual remains of the ancient organisms, such as bones and shells, or the evidence of their presence, such as trails and tracks. Their abundance and diversity are truly amazing. Some rocks (such as coal, chalk and certain limestone) are composed almost entirely of fossils, and others contain literally millions of specimens. Invertebrate marine forms are most common, but even large vertebrate fossils of mammals and reptiles are plentiful in many formations. For example, it is estimated that 50,000 fossil mammoths have been discovered in Siberia, and many more remain buried.

The Principle of Cross-cutting Relations: The relative age of certain geological events is also shown by the principles of the cross-cutting relations, which state that the igneous intrusions and faults are younger than the rocks they cut. Cross-cutting relations can be complex, however, and careful observations may be required to establish the correct sequence of events. The scale of cross-cutting feature is highly variable, ranging from large faults, with displacements of hundreds of kilometres to small fractures, less than a millimetre long.

The Principle of Inclusion: The principle of inclusion states that a fragment of rock incorporated or included in another is older than the host rock. The relative age of intrusive igneous rocks (with respect to surrounding rock) is therefore commonly apparent in inclusions. As a magma moves upward through the crust, it dislodges and engulfs large fragments of surrounding materials, which remain as unmelted foreign inclusion.

The principle of inclusion can also be applied to conglomerates in which relatively large pebbles and boulders eroded from pre-existing rocks have been transported and deposited in a new formation. The conglomerate is obviously younger than the formations from which pebbles and cobbles were derived. In areas where superposition or other methods do not indicate relative ages, a limit to the age of a conglomerate can be determined from the rock formation represented in its pebbles and cobbles.

Energy and Heat

The sun is the ultimate source of energy for the processes of change on the earth's surface and in the atmosphere. Our sun is a ball of constantly churning gases that are heated by continuous nuclear reactions. It is about average in size compared to other stars, and it has a surface temperature of about 6,000°C (about 11,000°F). Like all objects, it emits energy in the form of electromagnetic radiation. The energy travels outward in straight lines, or rays, from the sun at a speed of about 1,86,000 km (30,000 mi) per second. At that rate, it takes the energy about $8\frac{1}{3}$ minutes to travel the 150 million km (93 million miles) from the sun to the earth.

Influence of Global Warming

In the opinion of some of the climatologists the temperature of the earth is increasing, mainly because of the human activities. Other scientists, however, opine that the warming of the last few decades could be part of natural climatic cycle.

Carbon dioxide, produced by human activities, is a major cause of concern in climate warming. This gas is released to the atmosphere in large quantities by fossil fuel burning. The greenhouse effect is caused by atmospheric absorption of long-wave radiation, largely by carbon dioxide and water vapour, that is emitted from the earth. Also of concern are other gases that are commonly in very small concentration – methane (CH_4), nitrous oxide (NO), ozone (O_3), and chloro fluoro carbons (CFC). These gases also absorb long-wave radiation and enhance the greenhouse effect, even though they are even less abundant than CO_2. Taken together with CO_2, they are referred to as greenhouse gases.

Prior to the 1779 Industrial Revolution, carbon dioxide concentration in the atmosphere was at a level slightly less than 300 parts per million (ppm), or about 3/100ths (0.003%) of a per cent by volume. During the last one hundred years or so, that amount has been substantially increased by fossil fuel burning. When fuels like fuelwood, coal, oil, or natural gas are burned, they yield water vapour and carbon dioxide. The release of water vapour does not present a problem, because a large amount of water vapour is normally present in the global atmosphere. But, because the normal amount of CO_2 is so small, fossil fuel burning has raised the level to about 350 ppm. This is a 22 per cent increase.

The rapid increase in CO_2 occurred after the Industrial Revolution. Though, the CO_2 has increased with the passage of time since 1860, after 1940 or so the level, however, remained nearly stable. But, after 1940, CO_2 began a rapid rise – so rapid that at the present rate of increase (about 4 per cent per year), the amount of CO_2 in the air will double by 2030. Even if worldwide fossil fuel combustion is cut to half, the prediction is that doubling will occur by 2050, only about half a century away.

Humans have tilted this balance by clearing land and burning the vegetation cover as new areas of forest are opened for development. This practise increases the amount of CO_2 in the air when agricultural land is allowed to return to its natural forest state. CO_2 is removed from the air by growing trees. At present, scientists calculate that forests are growing more rapidly than they are being destroyed in the mid-latitude regions of the Northern Hemisphere. This plant growth helps to counteract the build-up of CO_2 produced by fossil fuel burning. It may be outweighed by the tropical deforestation that is taking place in South America, Africa and South-East Asia.

Laboratory experiments have shown that plants exposed to increased concentration of CO_2 will grow faster and better. The faster they grow, the more CO_2 they can take in, which helps to reduce the amount in the atmosphere. However, scientists are unsure whether increased CO_2 will stimulate plant growth under natural conditions.

Another part of the cycle involves the oceans. The oceans' surface layer contains microscopic plant life that takes in carbon dioxide. The CO_2 in the ocean water initially comes from these microscopic floating plants. They die and their bodies sink to the ocean bottom. There they decompose and release CO_2, enriching waters near the ocean floor.

This CO_2 eventually returns to the surface through a system of global ocean current flows involving both bottom and surface currents. In this system, cold CO_2-rich bottom waters rise to the surface in the northernmost Pacific, where CO_2 is released. Meanwhile, warmer, CO_2-poor waters sink in the northernmost Atlantic. In fact, the ocean acts like a slow conveyor belt, moving CO_2 from the surface to ocean depths and releasing it again in a cycle lasting about 1,500 years.

At present, scientists estimate that ocean surface waters absorb more CO_2 than they release, owing to increased atmospheric levels of CO_2. Therefore, carbon dioxide may be accumulating in ocean depths. However, current studies of global climate computer simulation indicate that the oceans may not be as effective in removing excess CO_2 as scientists previously thought.

Although there is a great deal of uncertainty about the movements and build-up rate of excess CO_2 released to the atmosphere by fossil fuel burning, one thing is certain: Without conversion to alternative energy sources, fuel consumption will continue to release carbon dioxide, and it is only a matter of time before the earth will feel its impact.

Differences in Temperature

The temperature and environment of the large urban places differ from that of the rural settlements and countryside. In fact, it is well established that urban micro climates generally differ from those of nearby non-urban areas. The surface energy characteristics of urban areas possess unique properties similar to desert locations. Because more than 65 per cent of the world's population will be living in cities by the year 2010, the study of 'urban heat islands' and other specific environmental effects related to cities is an important topic for physical geographers. The following factors contribute to urban micro climates:

Urban Surfaces Typically are Metal, Glass, Asphalt, Concrete, or Stone: These city surfaces conduct upto three times more heat than wet sandy soil. The heat storage capacity of these materials also exceeds that of most natural surfaces making cities *heat islands.* During the day and evening, higher temperatures exist above these surfaces.

At night such surfaces rapidly radiate this stored heat to the atmosphere, producing minimum temperatures some 5°-8°C (9°-14°F) warmer in urban areas as compared to rural areas, especially on calm, clear nights. As a result, both the maximum and the minimum temperatures are higher than in non-urban areas, although the effect of heat island characteristic is more profound during night time cooling.

Urban Surfaces behave Differently from Natural Surfaces in Energy Balance: Albedo values are lower, leading to a higher net radiation value. However, urban surfaces expend more of that energy as sensible heat than non-urban areas; more than 70 per cent of the net radiation in an urban setting is spent in this way.

The Irregular Geometric Shapes of a Modern City Affect Radiation Pattern and Winds : Incoming insolation is caught in maze-like reflection and radiation 'canyons' which tend to trap energy for conduction into surface materials, thus easing temperatures. In natural settings, insolation is more readily reflected, stored in plants, or converted into latent heat through evaporation.

Buildings tend to interrupt wind flows, thereby diminishing heat loss through adjective (horizontal) movement. Winds average 25 per cent velocity in cities than in rural areas, although buildings create local turbulence and funneling effects. Wind flows can significantly reduce heat island effect by increasing turbulent heat loss. Thus, maximum heat island effect in most build-up areas of a city occur on calm, clear days and nights.

Human Activity Alters the Heat Characteristics of Cities: In Delhi, Kolkata, Mumbai, Chennai, Karachi, Dhaka, New York, Tokyo, Mexico City, Sao Paulo, Rio-de-Janeiro, Lagos, Cairo, London, etc., for example, during the summer months, production of electricity and use of fossil fuels release an amount of energy equivalent to 25-50 per cent of the arriving insolation. In the winter, urban generated heat is on the average 250 per cent greater than arriving insolation.

Urban Surfaces are Generally Sealed (built on and paved) so that Water cannot Reach the Soil: Central business districts are on the average 50 per cent sealed, whereas suburbs are 20 per cent sealed. Because urban precipitation cannot penetrate the soil, water run-off increases.

Urban areas respond much as a desert landscape: a storm may cause a flash flood over the hard, sparsely vegetated surfaces, only to be followed by a return to dry conditions a few hours later. Little of the net radiation is expended for evaporation on such a surface.

The Amount of Air Pollution, including Gases and Aerosols, is Greater in Urban Areas than in Comparable Natural Settings: Pollution actually increases reflecting (albedo) in the atmosphere above the city, thus reducing insolation reaching the ground. The pollution blanket also absorbs infrared radiation, radiating the heat downward. Every major city produces its own dust dome of air-borne pollution, which can be blown from the city in elongated plumes, depending on wind direction.

The increased particulates present in pollution act as condensation nuclei for water vapour. Convection created by the heat island lifts the air and particulates, producing increased cloud formation and the potential for increased precipitation. Although these precipitation effects of the urban heat island are difficult to isolate and prove, research suggests that urban stimulated precipitation occurs downwind from cities.

Average Differences in Climatic Elements between Urban and Rural Environments

Element	*Urban Compared to Rural Environs*
	Contaminants
Condensation nuclei	10 times more
Particulates	10 times more
	Radiation
Total on horizontal surface	0-20% less
Sunshine duration	5-15% less
	Precipitation
Amounts	5-15% more
Snowfall, downwind (lee) of city	10% more
Thunderstorms	10-15% more
	Temperature
Annual mean	0.5-3-0°C (0.9-5.4°F) warmer
Summer maxima	1.0-3°C (1.8-3.0°F) warmer

Conclusion: To summarise, the two major cycles of air temperature — daily and annual — are controlled by the cycles of insolation produced by the rotation and revolution of the earth. These cycles induce cycles of net radiation at the surface. When net radiation in the daily cycle is positive, air temperatures increase, and when negative, air temperatures decrease.

This principle applies for both daily and annual temperature cycles. Days are warm and nights are cool as net radiation goes from positive to negative. Summers are warm and winters are cold because average net radiation is high in the summer and low in the winter. Surface characteristics affect temperatures, too. Rural surfaces are generally moist and slow to heat, while urban surfaces are dry and absorb heat readily. This difference creates an urban heat island effect.

Air temperatures normally fall with altitude in the troposphere. At the tropopause, this decrease stops. In the stratosphere above, temperatures increase slightly with altitude. Air temperatures observed at mountain locations are lower with higher elevation, and day-night temperature differences increase with elevation.

Daily and annual temperature cycles are influenced by maritime or continental location. Ocean temperatures vary less than land temperatures because water heats more slowly and can both mix and evaporate freely. Maritime locations that receive oceanic air, therefore, show smaller ranges of daily and annual temperature.

Global temperature patterns for January and July show the effects of latitude and maritime-continental location. Equatorial temperatures vary little from season to season. Poleward temperatures decrease with latitude, and continental surfaces at high latitudes can become very cold in winter. Isotherms over continents swing widely north and south with seasons, while isotherms over oceans move through a much smaller range of latitude.

The global temperature changes from year to year. Within the last few decades, global temperatures have been increasing. Global temperatures are projected to rise significantly if we continue to release large quantities of greenhouse gases. Climatic warming could lead to a rise in sea level, threatening coastal populations, and also cause agricultural patterns and natural ecosystems to shifts in response to changing climatic boundaries. An increase in the number of extreme climatic events has been linked to the change in climate and could increase in the future. The reduction in release of greenhouse gases by world nations is the subject of ongoing international diplomatic activity.

Energy Transfer Process

Energy transfer processes, including radiation, maintain a balanced heat budget at the earth surface. They are represented in the equation: Rs= H + LE + A. Here, R, is the surface radiation balance (the difference between absorbed short wave radiation and the net upward long-wave radiation from the surface), H the net transfer of sensible heat between surface and atmosphere by conduction and turbulent exchange, L the latent heat of evaporation, E the rate of evaporation, and A the flux of heat between the surface and lower layers of soil or water. Energy which raises the temperature of a substance is known as *sensible heat,* that which gets transformed in the alteration of the physical state of matter (notably water) is *latent heat.*

Together the energy transfer processes effect complex exchanges between the earth's surface and the atmosphere. Local and regional energy budgets are kept in balance by transfer of addictive, sensible, and latent heat.

The energy budget concept applies equally to the entire globe, geographic regions, buildings, plants, animals, and even human body. Annual heat budgets for the continents and oceans are summarised below in table.

Annual Heat Budgets of the Continents and Oceans (kilolangleys per year)

Continent	*Net Radiation*	*Latent Heat Flux*	*Sensible Heat Flux*	*Subsurface Transport*
Africa	68	26	42	0
Antarctica	-11	0	-11	0
Asia	47	22	25	0
Australia	70	22	48	0
Europe	39	24	15	0
North America	40	23	17	0
South America	70	45	25	0
All land	49	25	24	0
Ocean				
Arctic	-4	5	-5	-4
Atlantic	82	72	8	2
Indian	85	77	7	1
Pacific	86	78	8	0

Average Temperature of Surface

In order to study the climates systematically and to ascertain the influence of climatic factors on the economy and society of a macro or micro region, several kinds of temperature values are required.

Probably, the most used basic temperature value is the daily mean from which monthly mean and annual average values can be derived.

In practise the daily mean is found by adding the 24-hour maximum to 24-hour minimum and dividing by 2, because comparatively few stations in the world take hourly or continuous temperature observations, which would afford a better basis for statistical mean. The difference between the highest and the lowest temperatures of the day is the diurnal (daily) range.

Ordinarily, observations of the maximum and the minimum thermometers are the bases of the daily mean temperature and the diurnal range. The mean monthly temperature is found by adding the daily means and dividing by the number of days in the month. Mean monthly values for the year indicate the annual march of temperature through the seasons.

The annual range is the difference between the mean temperatures of the warmest and the coldest months. For most stations in the Northern Hemisphere, the warmest month is July and the coldest month is January.

When corresponding temperature values for a number of years are averaged, a generalised value useful in climatic description is obtained. However, such averages tend to obscure the year-to-year variability of temperature and possible climatic fluctuations.

In the middle and high latitudes, the length of the frost-free season is commonly a part of the temperature record. The frost-free season is defined as the number of days during which temperatures are continuously above 0°C.

The mean frost-free season is the difference in days between the mean date of the last frost in spring and the mean date of the first frost in autumn. It is often regarded as an important consideration in agriculture, but there is actually no simple, direct relation between plant growth and freedom from frost.

Mean Surface Temperatures (0°C) at Selected Latitudes

Latitude	*January*	*July*
35°	10	36
40°	5	24
45°	-2	21
50°	-7	18
55°	-11	16
60°	-16	14
65°	-23	12
70°	-26	7
75°	-29	3
80°	-32	2
85°	-38	0
90°	-41	-1

It may be observed from the above table that the mean surface temperature declines steadily from 35°N towards the pole. The decrease in temperature is, however, remarkably high in the winter season (January) to that of the summer season (May).

Variability in Insolation

The amount of insolation received on any date at a place on earth is influenced by:

1. Solar radiation reaching the outer limits (480 km, 300 miles) of the atmosphere, which depends on:
   (a) energy output of the sun,
   (b) distance from the earth to the sun, and
   (c) interstellar dust in the solar system.
2. Transparency of the atmosphere
3. Duration of the daily sunlight period.
4. Angle at which the sun's noon rays strike the earth.

The solar radiation reaching the outer limits depends on the energy output of the sun which to a large extent is influenced by the number and frequency of sun spots (a whirling mass of gas just within the sun's atmosphere). Higher the number of sun spots, more is the heat emitted from the sun. The ellipticity of the earth's orbit around the sun also influences the insolation received on the earth. At the stage of *aphelion* (July 4) the earth is about 152 million km from the sun and at *perihelion* (January 3) about 147 million km from the sun. Consequently, the amount of solar radiation reaching the outer atmosphere varies by ±3.5 per cent from the mean. Other factors influencing the heat budget and temperature, however, largely override this difference. Similarly, the effect of extraterrestrial dust is believed to be negligible.

Transparency in Atmosphere

Transparency of the atmosphere has a more important bearing on the amount of insolation which reaches the earth's surface. The effect of dust, clouds, carbon dioxide, water vapour, etc., on the incoming solar radiation and earth temperatures is immense.

Transparency is also a function of latitude, for at middle and high latitudes the solar beam must penetrate the reflecting-absorbing atmosphere at a lower angle than in tropical latitudes. This effect varies with the seasons, being greater in winter when noon sun is lowest on the horizon.

The duration of daylight (the photo period) also varies with latitude and the seasons. Longer the photo period, the greater is the total possible insolation (Table). At the equator, day and night are always almost equal. In the polar regions the daily photo period reaches a maximum of 24 hours in summer and a minimum of zero hours in winter. At its summer solstice, under clear skies, a polar area may receive more radiation per 24-hour day than lower latitudes, although the net radiation used for heating is reduced because of the high albedo of ice and snow surfaces.

Longest Possible Duration of Insolation

Latitude	0°	17°	41°	49°	63°	66½°	90°
Daylight	12hr	13hr	15hr	16hr	20hr	24hr	6 months

The effect of the varying angle of the solar beam can be seen in the daily march of the sun across the sky. At or near solar noon the intensity of insolation at the earth surface is greatest, but in the morning hours, when the sun is at a low angle, intensity is reduced. The same principle applies to latitude and seasons. In winter and at high latitudes the sun's noon angle is low; in summer and at low latitudes it is more nearly vertical. The oblique rays of the low-angle sun are spread over a greater surface than perpendicular rays, therefore they produce less heating per unit area.

The angle at which solar radiation strikes the earth's surface also depends on terrain features. In the Northern Hemisphere, southern slopes receive a more direct solar beam, where as northern slopes may be entirely in the shade. The possible hours of direct sunshine during winter in a deep valley may be reduced to zero by surrounding hills.

It is clear from the above discussion that the world distribution of possible insolation at the surface is closely influenced by latitude. At the equator, the annual amount is about four times that at either of the poles. As the direct solar rays shift seasonally from one hemisphere to the other, the zone of maximum possible daily insolation moves with it. In tropical latitudes the amount of possible insolation is constantly great, and there is little variation with the seasons. But, in the annual journey, the sun passes over all places between the Tropic of Cancer and the Tropic of Capricorn twice, causing two maximums. In latitudes between 23½ and 66½ degrees, the maximum and the minimum periods of insolation occur shortly after the summer and winter solstices, respectively. Beyond the Arctic and Antarctic Circles the maximum insolation coincides with the summer solstice, but there is a period during which insolation is lacking. The length of the period increases towards the poles, where it is of six months duration.

It is interesting to note that the maximum annual values of temperature are at about 20 degrees latitude, where the drier air permits a greater proportion of the radiant energy to penetrate to surface levels. Cloudy regions receive less insolation at the surface than do areas with predominantly clear weather. In general, high plateaus and mountains are favoured by more effective insolation because of the relatively clearer and less dense air at high altitude. It is because of this factor that Leh and Kargil (Laddakh) and Lhasa (Tibet) receive more effective insolation.

Distributing Horizontally

The average global surface temperature is about 13°C but local averages vary widely. On maps, the horizontal distribution of temperature is generally shown by isotherms (lines joining the points of equal temperature). On weather maps of small areas the actual observed temperatures are used as a basis for drawing isotherms, but on continental or world maps mean temperatures are frequently reduced to sea level equivalents by adding 6°C for each 1,000 metres of elevation. This adjustment essentially eliminates the effect of altitude on temperature and thus facilitates the mapping of horizontal temperature differences.

The horizontal distribution of temperature is largely determined by latitude. The general decrease in temperatures from the equator towards the poles is one of the most fundamental and best known facts of climatology (Table). If the effect of latitude were the only controlling factor affecting net radiation, we would expect a world temperature map to have isotherms lying parallel to each other in the same fashion as parallels of latitudes. But such is not the actual case.

Influencing Forces

Thus, the horizontal distribution of temperature is the function of numerous other physical factors. The factors which influence the horizontal distribution of temperature on the surface of the earth are: i) latitude, ii) altitude, iii) cloud cover, iv) coastal and continental locations, v) ocean currents, and vi) nature of the surface.

Latitude: The latitudinal location of a place has a close influence on insolation which determines the temperature. In general, the lower the latitude of a place, the more the solar energy it receives annually. The sun's rays are vertical on equator and slanting on higher latitudes.

Moreover, when solar rays approach the earth at a more nearly vertical angle in lower latitudes, they must pass through a smaller thickness of absorbing and reflecting atmosphere before reaching the surface. Thus, the intensity of insolation decreases as one moves away from the sub-polar point, a point that migrates between the Tropic of Cancer and Tropic of Capricorn – that is, between 23.5°N and 23.5°S – during the year. In addition, daylight and sun angle change throughout the year, increasing the seasonal effect with increasing latitude. Consequently, from equator to poles, earth ranges from continually warm, to seasonally variable, to continually cold.

The effect of latitude on seasonal variation is also significant. For example, more solar energy is received at the poles at summer solstice than at the equator. Thus, the latitude and time of year have a close bearing on the horizontal distribution of temperature.

Altitude: In general, in the troposphere, temperatures decrease with increasing altitude above earth's surface. The normal lapse rate of temperature change with altitude in 6.4°C/1,000 metres or 3.5°F/1,000 feet. Thus, throughout the world, mountainous areas experience lower temperatures than do regions near sea level even at similar latitudes. The density of atmosphere also diminishes with increasing altitude. The density of atmosphere at an elevation of 5,500 metres (18,000 feet) is about half of that at sea level. As the atmosphere thins, its ability to absorb and radiate heat is reduced.

The consequences are that average air temperatures at higher elevations are lower, night-time cooling increases, and the temperature range between day and night and between areas of sunlight and shadow is also greater. Temperatures may decrease noticeably in the shadows and shortly after sunset. Surfaces both gain heat rapidly and lose heat rapidly to the thinner atmosphere. Also, at higher elevations the insolation received is more intense because of the reduced mass of atmospheric gases. As a result of this intensity, the ultraviolet energy component causes sunburn and pneumonia (a distinct hazard). It is because of this factor that tourists at high altitudes like Kargil, Leh (Laddakh) and Lhasa (Tibet) are always advised to get acclimatised for a few days before going for any outdoor adventure.

The existence and oscillation of the snowline in the mountains is mainly because the temperature decreases as the altitude increases. In fact, the snowline's location is a function both of elevation and latitudes if the elevation is great enough. In equatorial mountains, the snowline occurs approximately 5,000 metres (16,400 feet) and permanent ice fields and glaciers exist on equatorial mountain summits in the Andes and east Africa. With increasing latitude towards the poles, snowline gradually descents in elevation from 2,700 metres (8,850 feet) in mid-latitudes to lower than 900 metres (2,950 feet) in southern Greenland.

Two cities in Bolivia illustrate the interaction of the two temperature controls — latitude and altitude shows temperature data for the cities of Concepcion and La Paz, which are almost near the same latitude (about 16°S). Note the elevation, average annual temperature and precipitation for each city (Table). The hot humid climate of Concepcion at its much lower elevation (490 metres) stands in marked contrast to the cool, dry climate of highland La Paz. People living around La Paz actually grow wheat, barley, and potato — crops characteristically grown in cooler mid-latitudes — despite the fact that La Paz is 4,103 metres (13,461 feet) above the sea level. The combination of elevation and equatorial location guarantee La Paz nearly constant daylight and moderate temperatures, averaging about 9°C (48°F) for every month. Such moderate temperature and moisture conditions lead to the formation of more fertile soils than those found in the warmer and wetter climate of Concepcion.

Station

	Concepcion (Bolivia)	*La Paz (Bolivia)*
Latitude/longitude	16°15′S, 62°03′W	16°30′S, 68°10′W
Elevation	490 metres (1607.6 feet)	4103 metres (13,461 feet)
Average annual temperature	24°C (75.2°F)	9°C (48.2°F)
Annual range of temperature	5°C (9°F)	3°C (5.4°F)
Annual precipitation	121.2 cm (47.7 in)	55.5 cm (21.9 in)
Population	768,000	993,000

Cloud Cover: The distributional pattern of temperature is also significantly influenced by the type, height, and density of cloud cover. Orbital surveys reveal that approximately 50 per cent of the earth is cloud covered at any one time. Clouds are the most variable factor influencing earth's radiation budget, and they are the subject of much investigation and effort to improve computer models of atmospheric behaviour and weather forecasting.

Clouds have moderating influence on temperature, producing lower daily maximums and high night time minimums. Acting as insulation, clouds hold heat energy below them at night, preventing more rapid radiation losses, whereas during the day, clouds reflect insolation as a result of their high albedo values. The moisture in clouds both absorbs and liberates large amounts of heat energy, yet another factor in moderating temperatures at the surface. The earth atmosphere energy system responds with slightly low temperatures as a result of cloud cover.

Coastal and Continental Locations: The location of a place in relation to sea/ocean and interior of a continent also controls the distributional pattern of temperature. The physical nature of the substances themselves – rock and soil vs. water – is the reason for the land-water heating differences. More moderate temperature patterns are associated with water bodies, compared to more extreme temperatures inland. The coastal towns and cities have more uniform temperatures throughout the year as they receive marine air from the prevailing winds. These places remain cooler in summer and warmer in winter. Contrary to this, the town and cities located in the interior parts of the continents show much larger annual variations in temperature. In general, the continents tend to become warmer in summer than oceans in the same latitudes but appreciably colder in winter; the larger the landmass, the greater the contrast. Thus, extreme temperatures with a large seasonal range are the results of continentality, which affects the Northern Hemisphere more than the Southern Hemisphere.

Evaporation: More of the energy arriving at the ocean's surface is expended for evaporation than is expected over a comparable area of land. An estimated 84 per cent of all evaporation on the earth is from the oceans. When water evaporates and thus changes to water vapour, heat energy is absorbed in the process and is stored in the water vapour. This stored heat energy is called *latent heat.* One can experience this evaporation heat loss (cooling) by wetting the back of your hand and then blowing on the moist skin. Sensible heat energy is drawn from your skin to supply some of the energy for the water's evaporation, and you feel the cooling. Similarly, as surface water evaporates, substantial energy is absorbed, resulting in a lowering of nearby air temperatures. The land containing far less water, experiences far less evaporation, and therefore is moderated less by evaporative cooling.

Transparency: The transmission of light differs between soil and water; solid ground is opaque and water is transparent. Consequently, light striking a soil surface does not penetrate and gets accumulated during times of exposure and is rapidly lost at night or in shadow. The maximum and the minimum temperatures are generally experienced right at ground level. Below the surface, even at shallow depths, temperatures remain about the same throughout the day. This situation often exists at a beach, where surface sand may be painfully hot to feet but the sand a few centimetres below the surface feels cooler and offers relief.

In contrast, when light reaches a body of water, it penetrates the surface because of water's transparency, transmitting light to an average depth of 60 metres (200 feet) in the ocean. This illuminated zone is known as the *photic layer* and has been recorded in some ocean waters to a depth of 300 metres (1,000 feet). This characteristic of water results in the distribution of available heat energy over a much greater depth and volume, forming a larger heat reservoir than that made up of the surface layers of the land.

Specific Heat: When equal volumes of water and land are compared, water requires far more heat to raise its temperature than does land. In other words, water can hold more heat than can soil or rock, and therefore, water is said to have a higher specific heat. A given volume of water represents a more substantial heat reservoir than an equal volume of land, so that changing the temperature of the oceanic heat reservoir is a slower process than changing the temperature of land.

Movement: Land is rigid, solid material, whereas water is a fluid and is capable of movement. Differing temperature and currents result in a mixing of cooler and warmer waters, and that mixing spreads the available heat over an even greater volume than if the water were still. Surface water and deeper water mix, redistributing heat energy. Both ocean and land surfaces radiate heat at night, but land loses its heat energy more rapidly than does the moving mass of the oceanic heat reservoir.

Ocean Currents: Ocean currents are either warm or cold water in nature. They keep coastal water somewhat warmer or cooler than expected. As a specific example, the Gulf Stream moves northward off the east coast of North America, carrying warm water far into the North Atlantic Ocean. As a result, the southern coast of Iceland experiences much milder temperatures than would be expected for a latitude of 65°N, just below the Arctic Circle (66.5°N). In Reykjavik, on the southwestern coast of Iceland, monthly temperatures average above freezing during all months of the year. The Gulf Stream affects Scandinavia and northwestern Europe in the same manner.

In the western Pacific Ocean, the Kuroshio current, similar to Gulf Stream functions much the same in its warming effect on Japan and the Aleutian Islands. In contrast, along mid-latitude west coasts, cool ocean currents moderate air temperatures. The Canary current (west coast of Africa), Benguela current (the western coast of Namibia), California current (west of California), and the Peru current (west of Chile and Peru) are some of the important cold water currents which make the ocean's water cool even in the subtropical regions.

Nature of Surface

The earth surface is covered by land and water. The land surface has different types of rocks. Similarly, the proportion of salinity varies in the oceans. Land surfaces, in general, heat up more rapidly and intensely than the water surfaces. On the other hand, land cools more rapidly when the source of heating is cut off.

Moreover, solid earth is a bad conductor of heat, hence a shallow layer is more intensely heated. The earth is opaque, while water allows the rays to penetrate to a greater depth and so affect more water to a lesser extent. The daily fluctuations in temperature may be detected 15-18 metres (50-60 feet) below the surface of water, but not more than a metre below the surface of the ground. A water surface allows evaporation, and transference through mixing by convection currents readily set up in water. A piece of coloured paper left on the surface of a snow field will in a few hours have sunk to a depth of several centimetres because it has absorbed heat and so melted the snow around. A snow field reflects a large proportion of the rays, so that the snow remains while skiers become bronzed and get sun burns in the brilliant sunshine.

The sandy soils have a low specific heat and warm up rapidly at the surface. Swamps and waterlogged soils act like a water surface, and forests modify heating by casting shade.

Pattern of Distribution

Temperatures Decrease from the Equator to the Poles: In general, the annual insolation decreases from the equator to the poles, thus causing temperatures to decrease. In the month of January, the average temperature in the equatorial region is 25°C while on the North and South Poles the average temperatures read -35°C and -30°C, respectively.

Extreme Low Temperatures are found in the Arctic and Sub-Arctic Regions: Eurasia and North America are the largest landmasses in the world. The lowest temperatures in the month of January are recorded in these areas. For example, in the month of January Verkhoyansk (Yana basin, Siberia) records mean monthly temperature of -50°C (-58°F). The cold centre over northern Canada is also quite cold (-35°C, -31°F). Greenland also records -40°C temperature in January. An important factor in keeping winter temperatures low in these regions is the high albedo of snow cover, which reflects much of the winter insolation back to space.

The Annual Range of Temperature is Low in the Equatorial Region: The sun's rays are generally vertical over the equatorial region. The January and July temperatures range between 25°C and 30°C, respectively. The low range and uniformity of temperatures in the equatorial region are primarily because insolation at the equator does not change greatly with the seasons.

Isotherms make a Large North-South Shift from January to July in the Mid-latitudes in Northern Hemisphere: As we observed from the previous study that the isotherms in the months of January and July have a north-south shift, respectively. In the winter, isotherms dip equator ward, while in the summer they arch poleward. For example, at Verkhoyansk (Siberia), the mean January temperature reads -50°C, while in July it records about 15°C as the mean monthly temperature. The striking difference is due to the contrast between oceanic and continental surface properties, which cause continents to heat and cool more rapidly than oceans.

Highlands are Always Colder than Surrounding Lowlands: The Himalayan mountain ranges are much colder in both January and July than the Gangetic plain. In January and July, the Himalayan region records about 0°C and 20°C temperatures, respectively, while the corresponding temperatures in the Gangetic plain (Allahabad) read about 20°C and 35°C, respectively. The principle at work here is that temperatures decrease with an increase in altitude.

Records of Extreme Temperatures

Record	°C	*Location*	*Date*
Highest official air temperature	58	Azizia, Libya	Sept. 13, 1922
Highest US temperature	57	Greenland Ranch (Death Valley), California	July 10, 1913
Highest mean annual temperature	31	Lugn, Somalia	13-year mean
Lowest official temperature in Northern Hemisphere	-68	Verkhoyansk, Siberia	Feb. 5 and 7, 1892
Lowest official temperature in Western Hemisphere	-66	On Greenland ice cap at 2990 m	Dec. 6, 1949
Lowest temperature on North American continent	-63	Snag, Yukon	Feb. 3,1947
Lowest US temperature	-62.1	Prospect Creek Camp, Alaska	Jan. 23, 1971
Lowest record by US observers	-80.6	Amundsen-Scott Station (90°S)	July 22,1965
Lowest world surface air temperature	-88.3	Vostok Soviet Station (78°27′S 106°52′E at 3420 m)	Aug. 24, 1960

Areas of Perpetual Ice and Snow are Always Intensely Cold: The largest ice sheets of the world are found in Antarctica and Greenland. The perpetual snow reflects back over 80 per cent of the insolation to space in the form of albedo. Moreover, the surfaces of Greenland and Antarctica are over 3,000 metres (about 10,000 feet) in their centres. Since little solar energy is absorbed, little is available to warm the snow surface and the air above it. The Arctic Ocean, bearing a cover of floating ice, also maintains its cold temperatures throughout the year. However, the cold is much intense in January than on the Greenland ice sheet, since ocean water underneath the ice acts as a heat reservoir.

Temperature: Annual Range

The difference between the hottest and the coldest month is known as the *annual range of temperature.* The annual range of temperature varies from region to region and from place to place.

Chief Characteristics

The salient features of the annual range of temperature are as follows:

1. The annual range of temperature increases with latitude, especially over Northern Hemisphere. This trend is most clearly shown for Eurasia and North America. This is due to the contrast between summer and winter insolation, which increases with latitude.
2. The greatest annual range of temperature occurs in the sub-Arctic and Arctic zones of Asia and North America. There are two very strong centres of large annual range — one in the northeast of Siberia and the other in the northwest Canada-eastern Alaska. In these regions, summer insolation is nearly the same as that at the equator, while winter insolation is very low.
3. The annual range is moderately large on land areas in the tropical zone, near the tropics of Cancer and Capricorn. These are the regions of large deserts, e.g., Sahara (North Africa), Rub-al-khali (Saudi Arabia), Thar desert of India and Pakistan, Kalahari (South Africa), and Gibson (Australia). Dry air and the absence of clouds and moisture allow these continental locations to cool strongly in winter and warm strongly in summer, even though insolation contrasts with the season are not as great as at higher latitudes.
4. The annual range over oceans is less than that over land at the same latitude. It can be seen that at 40°N latitude, starting from the right, the range of temperature is between 5°C and

10°C (9°F and 18°F) over the Atlantic Ocean, but increases to about 30°C (54°F) in the interior North America. In the Pacific, the range falls to 5°C (9°F) just off California coast and increases to 15°C (27°F) near Japan.

In Central Asia, the annual range of temperature is near 35°C (63°F). Again, these major differences are due to the contrast between land and water surfaces. Since water heats and cools much more slowly than land, a narrow annual range in temperature is experienced.

5. The annual range of temperature is very small over oceans in the tropical zone. It may be seen that the annual range of temperature is less than 3°C (5°F) over the equatorial water. It is mainly because the water heats and cools slowly.

Diurnal (Daily) Range: The amount of variation between the maximum and minimum of any element, such as air temperature (and relative humidity), during 24 hours is known as *diurnal range.* The diurnal range of temperature is usually greatest in desert regions which record high daytime temperatures followed by a rapid heat loss through *radiation* at night, owing to the clear skies.

Frost-free Season: From the agricultural and economic point of view, the frost-free season has great significance. The *frost-free season* is defined as the number of days during which temperatures are continuously above 0°C. The mean frost-free season is the difference in days between the mean date of the last frost in spring and the mean date of the first frost in autumn.

Inversion of Temperature: Normally air temperature decreases with increasing elevation but under certain weather conditions the converse may be true. Thus, contrary to the normal environmental lapse rate, over a limited height range, air temperature increases with height so that a layer of warmer air overlies a coider layer. A surface inversion is commonly experienced in valleys and hallows. The inversion of temperature takes place:

1. In winter on calm, clear nights.
2. When radiation has caused considerable cooling and the cold air has sunk down into them.

3. When the sky is clear and anticyclone condition prevails.
4. When the earth surface is covered with ice, snow and frost.

There are two types of inversion of temperature:

A High-altitude Inversion: The high-altitude inversion occurs mainly due to the frontal convergence, when a warm air mass is forced from the ground surface by the underlying cold air mass at a cold front. Alternatively, a similar inversion can be created when a warm air mass overrides a colder one along a warm front. Upper-air inversion also develops in the sub-tropics associated with the deep subsidence and adiabatic warming in a warm anticyclone.

A Surface Inversion: A surface inversion is much more localised and is often dependent on the terrain. It frequently occurs during winter anti-cyclonic weather when, during calm, cloudless nights, there is a rapid heat loss from the ground by radiation. Some mountain settlements have been sited to avoid these cold 'spots'.

The inversion of temperature adversely affects orchards, fruit trees and vegetables. In countries like France, Italy, Spain, and USA oil-burning heaters are used to warm the surface layer and create air circulation or large fans are used to mix the cool air at the surface with the warmer air above, in the orchards and vegetable fields.

Low-level temperature inversions often occur over snow-covered surface in winter. Inversions of this type are very intense and can extend thousands of metres into the air. They build up over many long nights in the Arctic and polar regions, where the solar heat of the short winter day cannot completely compensate for night time cooling.

Distributing Vertically

The permanent snow caps on high mountains, even in the tropics, indicate the decrease of temperature with altitude. The average rate of temperature decrease upward in the troposphere is about 6.4°C per km extending to the tropopause. This vertical gradient of temperature is commonly known as the *standard atmosphere lapse rate* or *normal lapse rate*, but it varies with height, season, latitude and other factors. The *normal lapse rate* represents the average of many observations of vertical temperature distribution and should not be confused with the *actual lapse rate*, which indicates temperature values above a given location at a given time. Indeed, the actual lapse rate of temperature does not always show a decrease with altitude. Where observations indicate no change with altitude, the lapse rate is termed as *isothermal*. Such a condition never occurs over a very great vertical range nor for a long period of time in the troposphere.

The Sun

As solar radiation travels through space, none of it is lost. However, the rays spread apart as they move away from the sun. This means that a planet farther from the sun, like Mars, receives less radiation than one located near the sun, like Mercury and Venus. The earth intercepts only about one half of one billionth of the sun's total energy output.

Warming the Atmosphere

The atmosphere is not directly heated by the sun from above. It is rather heated from below. The processes of heating of the atmosphere are: *convection, conduction,* and *radiation*.

Convection: The transmission of heat from one part of a liquid or gas to another by movement of the particles themselves is known as *convection*. When the lower part of a mass of liquid is heated, it expands, its density is reduced, and it rises, carrying its heat with it — to be replaced by cool fluid which in turn is heated. A familiar example of convection is the upward movement of air which has been heated by contact with the earth's surface; this air is said to rise in a convection current. In other words, convection requires a mobile medium like air or water; it is readily observed when a dish of water is heated on a stove. The hot water at the bottom of the pan rises and circulates to redistribute the heat. A bubbling pot is an example of convection at work.

Conduction: Conduction is a process in which heat is transferred directly through matter from a point of high temperature to a point of low temperature by molecular impact but without overall movement of the matter itself. Conduction transfers heat between adjacent molecules. Heat passes from warmer to colder substances as long as temperature difference exits. Thus, when the surface absorbs radiation and warms above the air temperature, conduction transfers part of the heat to the lower layer of air.

When the surface is cooler than the overlying air, the heat transfer is reversed and the air is cooled. The latter phenomenon is common over land at night and during winter in the middle and high altitudes. Air is a poor conductor of heat.

Radiation: It is a process by which a body emits radiant energy (energy received from the sun), e.g., in the form of heat. It causes a loss of heat and, therefore, leads to cooling. Although the sun is the initial and primary source of the heat, the air is heated mainly by reradiation from the underlying land or water surface. Radiant energy is constantly emitted in all directions by the sun. Some of this energy reaching the earth is being converted into heat. Earth is constantly losing heat into space by radiation. During the day, the heat received from the sun by insolation exceeds the amount lost by radiation, and the temperature rises till a maximum is reached. At night, the reverse is the case, and the temperature falls till a minimum is reached. In summer, when the days are long and the sun is high, the ground grows gradually warmer, but in winter when the days are short, and the sun is low, radiation so far exceeds insolation that the ground becomes cold. The land loses radiation more rapidly than water, and high ground more rapidly than low ground.

Radiation of Sun

Solar radiation that is intercepted by the earth is known as *insolation.* The amount of insolation reaching the outer limit (480 km, 300 miles) of the atmosphere is termed as *solar constant.* The amount of insolation intercepted by earth on a surface perpendicular to the sun's rays when earth is at its average distance from the sun is 1370 watts per square metre, averaged over the entire globe at the thermopause or 1.968 calories per sq cm per minute. Insolation is measured with the help of pyranometers.

The amount of insolation reaching any place during one day depends on:

1. The area and nature of the surface,
2. The inclination of the rays of the sun,
3. The transparency of the atmosphere, and
4. The position of the earth on its orbit.

The amount of solar constant varies throughout the day with the changes in the angle of incidence of the sun's rays; and the length of time that the sun remains above the horizon is a determining factor. At the equinoxes, that time is twelve hours per day everywhere on the equator, but towards the summer solstice the longer days of the higher latitudes more than compensate for the great obliqueness of the sun's rays. Thus, the total insolation per day is not at its maximum at the latitude where the sun is highest at noon, but at a rather higher latitude where the daily duration of the sunlight is greater. For the year as a whole, however, insolation is greatest at the equator, it decreases at first slowly then more rapidly, and slowly again towards the poles. Insolation shows the least variation throughout the year at the equator, but varies very considerably at the poles.

Insolation is the single energy input driving the earth's atmosphere system. It includes all radiation arriving at earth's surface, both direct and diffuse (as downward scattered).

For example, insolation decreases poleward from 25 degrees latitude in both the northern and the southern hemispheres. Consistent daylight and high sun altitude produce average insolation of 180-200 W/sq m throughout the equatorial and tropical latitudes. In general, insolation of 240-280 W/sq m occurs in low-latitude deserts worldwide because of frequently cloudless skies. Note the energy pattern in cloudless subtropical deserts in both hemispheres (for example, the Sonora, Saharan, Arabian, Atacama, Namibia, Kalahari, and Australian-Gibson deserts).

Albedo: The proportion of solar radiation falling on a non-luminous body which the latter reflects, usually expressed as decimal, is known as *albedo*. In other words, it is the reflective quality of a surface, expressed as the relationship of incoming to reflected insolation and stated as percentage, a function of surface colour, angle of incidence, and surface texture.

The albedo of the earth is approximately 0.4, i.e., about 40 per cent of the solar radiation is reflected back into space. The value is much higher for a snow covered surface, and lower for dark soil. Albedo varies within wide limits, depending on the substance involved.

In general, darker colours have lower albedos, and lighter colours have higher albedos. On water surface, the angle of the solar rays also affects albedo values; lower angles produce a greater reflection than do higher angles. In addition, smooth surfaces increase albedo, whereas rougher surfaces reduce it. Specific locations experience highly variable albedo during the year in response to changes in cloud and ground cover. The rate of albedo is between 19 and 38 per cent between the tropics (23.5°N to 23.5°S), to as high as 80 per cent in the polar regions. In general, the earth's atmosphere system directly returns back to space slightly less than one-third (29 to 34 per cent) of the solar radiation it receives.

Considering normal amounts of cloud and snow cover and other factors, the albedo of the earth, as a whole, has been estimated at about 40 per cent (recent observations by satellites seem to indicate somewhat lower values), but further measurements are needed to determine long-time global averages.

Comparison between Heat Budget and Radiation

Radiation is the means by which solar radiation reaches the earth and the earth loses energy to outer space. The global radiation budget has three major components:

- Solar radiation incoming at the outer limits of the atmosphere (Qs),
- The planetary albedo (a), and
- Outgoing long-wave radiation from the earth to space (I).

Thus, the basic form of the budget equation for the earth and its atmosphere is: R = Qs (1-a)-I. Here, R is the radiation balance (surplus or deficit) and (1-a) in the percentage of total insolation which is absorbed by the earth and atmosphere. Note that whereas insolation is intercepted on an area equivalent to the cross section of the earth atmosphere, terrestial radiation is emitted from the entire sphere, an area four times larger. Of the total incoming (Qs) about 26 per cent is reflected by clouds or scattered back to space by clouds, dust, and gas molecules without heating the air; 4 per cent is reflected to space from the earth's surface. The earth's mean albedo (known as the *planetary albedo)* is about 30 per cent, although its value is greater in polar regions than at the equator, varying with the angle of incidence as well as the characteristics of the reflecting surfaces. The albedo of clouds varies widely with their thickness and composition. On the average, clouds cover about half of the globe, giving them a major influence on the radiation budget (Table).

Albedo of Various Surfaces

Surface	*Per cent Reflected*
Clouds, stratus	
< 150 metres thick	25-63
150-300 metres thick	45-75
300-600 metres thick	59-84
Average of all types and thicknesses	50-55
Concrete	17-27
Crops, green	5-25
Forest, green	5-10
Meadows, green	5-25
Ploughed field, moist	14-17
Road, blacktop	5-10
Sand, white	30-60
Snow, fresh-fallen	80-90
Snow, old	45-70
Soil, dark	5-15
Soil, light (or desert)	25-30
Water	8*

* Typical value for water surface, but the reflectivity increases sharply from less then 5 per cent when the sun's altitude above the horizon is greater then 30 degrees to more then 60 per cent when the altitude is less then 3 degrees. Rough seas have a somewhat lower albedo than calm seas.

About 19 per cent of insolation is absorbed in the atmosphere by gases, clouds and suspended solids. Oxygen and ozone at levels absorb most of the ultraviolet radiation to provide the main source of energy for circulation above 30 km. Water vapour, clouds, and dust are the principal absorbers of incoming long-wave radiation in the troposphere. The earth's surface absorbs 51 per cent of insolation, either directly or after diffuse scattering downward by clouds and atmosphere. Thus, approximately 70 per cent of total insolation is effective in heating the earth and its atmosphere

The earth is also a radiating body. In contrast to sun, which radiates at a temperature of 6,000°K, the earth has an average temperature of only 288°K and emits terrestrial radiation at much greater wavelengths. Although the constituents of the atmosphere collectively absorb only about one-fifty of the incoming short-wave radiation, they can capture a large part of the outgoing long-wave radiation. This ability to admit most of the insolation yet retard losses by radiation from the earth's surface is commonly known as the *greenhouse effect;* a more appropriate term is *atmosphere effect.* Besides the long-wave radiation from the earth's surface, there is also radiation from cloud layers, gases (especially water vapour and carbon dioxide) and dust to space. The total amount of energy reaching the earth over a considerable period of time is equalled to total outward losses. If this were not so the earth would soon become either very hot or very cold. Satellite observations have confirmed a fairly stable global energy budget over relatively long periods.

The radiation budget of a particular place is, however, seldom in balance. Actually, there is an annual deficit at high latitudes and a surplus at low latitudes owing to differences in the angle of incidence and surface albedos. At any latitude variations in slope, exposure, and the character of surface cover (for example, water, land or vegetation) produce regional differences in the radiation budget. In the middle and low latitudes a true radiative balance may exist briefly near the beginning and end of each daylight period, when deficits and surpluses replace one another.

Airwaves and Waves

The movement of air in the atmosphere is known as *atmospheric circulation.* The circulation of atmosphere is a complex issue which has not been fully explored so far. The meteorologists have, however, discovered certain universal laws which explain the broad aspects of large-scale motion near the surface of the earth. Some of the important laws of atmospheric circulation have been described in the paragraphs that follow.

Role of Trade Winds

The *trade winds* are the surface winds of the Hadley cells as they move from the horse latitudes to the doldrums. In the Northern Hemisphere, they are the North-East Trades, and in the Southern Hemisphere the South-East Trades. It may be mentioned again that winds are named by the direction from which they blow. A west wind blows from the west towards the east; a northeast wind blows from the northeast towards the southwest.

The trade winds blow with great regularity throughout the year over the oceans. They derive their name from the Latin *trado,* meaning constant direction, and has given rise to the phrase *'to blow trade'*, meaning *'to blow along a regular track'*. In continental interiors they blow much less steadily than over the oceans, but are fairly regular over the eastern side of the oceans than over the western sides, where there is some interference from pressure disturbances on the windward shores. As a result, the eastern coastlines and offshore islands in the trade wind belt exhibit very arid conditions (e.g., the Canary Islands) owing to the dryness of the air mass and the strong low level inversion of temperature, which prevents vertical cloud development. Conversely, on windward coasts, the moisture-laden air mass and the much weaker temperature inversion see stronger convection developed with accompanying rainstorms.

The main function of the trade winds is to remove surplus heat from the subtropical high pressure belt by evaporating great quantities of heat from the tropical oceans, thereby helping to maintain the global heat balance as they return surplus air to the equatorial zone. With the seasonal change in the declination of the sun, the trade winds also move northwards and southwards, their range averaging about 5 degrees latitude. Although weather in the trade wind region is normally fine and quite, tropical cyclones are often experienced here. In general, the average speed of trade winds varies between 16 and 30 km (10 and 20 miles) per hour, while calms are extremely rare and clear sunny conditions are usual.

Between the two trade wind systems is found what is known as the *Intertropical Convergence Zone* (ITCZ); part of it is occupied by an area of calms or light winds – the *doldrums*. However, it has been recently discovered that in the 'summer' in the Northern Hemisphere, and especially over the continents, a zone of westerly winds intervenes between the 'trades' which are called the *equatorial westerlies.*

Role of Anti-trade Winds

The westerly winds, which blow with great frequency and regularity in regions lying on the poleward sides of the subtropical high pressure areas or horse latitudes, are known as the *anti-trade winds* or *westerlies.* The areas in which they blow are often known as the *regions of the westerlies.* The prevailing wind direction of the anti-trade winds in the Northern Hemisphere is from the southwest to the northeast and in the Southern Hemisphere from the northwest to the southeast. In winter they move southward in the Northern

Hemisphere, affecting the Mediterranean regions and much of the horse latitudes, bringing winter rain to those areas. In summer they move northward again, between 40°N to the Arctic Circle (66½°N) and 35°S to the Antarctic Circle (66½°S). They, however, blow throughout the year.

These winds derive their name from the prevailing direction. They blow opposite to the direction of trade winds and are thus known as the *anti-trade winds.* They are very variable in force and direction, especially in the Northern Hemisphere, where the winds may be easterly for a considerable period.

The weather in these areas is marked by an almost constant procession of depressions and anticyclones moving eastwards, and is made more complex in the Northern Hemisphere by the alternation of oceans and continents. On the eastern side of Asia, for instance, the effect of the westerlies is broken by the monsoon. In winter, the North Atlantic Ocean is probably the stormiest region of the world at any season, but in summer the winds and weather are much less violent. In the Southern Hemisphere, on the other hand, the westerlies blow with great strength and regularity throughout the year over the almost unbroken expanse of ocean, and have given the name *roaring forties* to the region.

Roaring Forties: The region between latitudes 40°S and 50°S where the prevailing westerly winds, being unobstructed by land, blow over the open oceans with great regularity and strength. The frequency of depressions brings storminess, cloudiness, and rough seas – a notoriously hazardous shipping zone.

Arctic Easterly Winds

A belt of easterly winds blowing in a zone between the westerly depression tracks and the polar high pressure belts. They are generally sporadic and of low velocity. From the polar areas cold air tends to move equatorward. This movement is pronounced in the Southern Hemisphere, where both the high over the Antarctic plateau and the subpolar low over the southern ocean are clearly defined. As the air moves equatorward it is deflected to the left, and so a belt of winds results, spiraling from the polar region in an easterly direction, forming a polar vortex. In the Northern Hemisphere, pressure and wind conditions are so complicated that these polar winds are extremely irregular. Sometimes a large part of North America and/or Asia is affected by the polar winds which spread southward. These winds are extremely cold. In general, in the Northern Hemisphere, they blow from northeast to southwest, whereas in the Southern Hemisphere, from the southeast to the northwest.

Upper Atmospheric Circulation: Middle and upper atmospheric circulation is an important component of the atmosphere's general circulation. These upper atmosphere winds tend to blow from west to east from the subtropics to the poles.

Within the westerly flow of geostrophic winds are great waving undulations called *Rossby waves,* named after meteorologist C.G. Rossby, who first described them mathematically. The polar front is the contact between colder air to the north and warmer air to the south. The Rossby waves bring tongues of cold air southward, with warmer tropical air moving northward. As these disturbances mature, distinct cyclonic circulations form, with warmer air and colder air mixing along distinct fronts. The development of cyclonic storm systems at the surface is supported by these wave-and-eddy formulations and other upper air flows. These Rossby waves develop along the flow axis of a jet stream.

Shifting of Seasons

The earth revolves on its orbit round the sun in 365 days. The earth is inclined on its axis at an angle of 23½ degrees. Thus, the relative position of the earth with the sun changes. In consequence, the positions of all the pressure belts change with the northward and southward migration of the sun. At the time of summer solstice the sun is vertical over the Tropic of Cancer (21st June) and therefore all the pressure belts except the northern polar high pressure belt shift northward. In the Northern Hemisphere, the equatorial low pressure belt prevails between 0° latitude (equator) and 10°N latitudes, the subtropical belt extends between 30° and 40°N latitudes. Thus, all the wind belts associated with the said pressure belts also shift northward. The sun becomes vertical over the equator at the time of autumnal equinox (23rd September) and hence all the pressure belts which shifted to the north occupy their normal positions. After autumn equinox, there is southward migration of the sun, which becomes vertical over the Tropic of Capricorn at the time of winter solstice (23rd December) and hence the pressure and wind belts shift southward except the southern polar high pressure belt. Thereafter, the sun again becomes vertical over the equator at the time of spring (vernal) equinox (21st March) and hence all the pressure and wind belts occupy their normal positions. Thus, there is shifting in the position of pressure and wind belts due to seasonal changes of position of the earth in relation to the sun. These seasonal changes in the relative positions of the pressure and wind belts influence the areas of the Mediterranean climate and the position of the westerlies which move northward in the summer season (June) and southward in the winter (December) season. The summer and winter monsoons and the complete reversal of wind direction are also the result of shift of wind and pressure belts. Both the trade and anti-trade winds are thus closely influenced by the shifts in pressure belts.

Jet Stream: Types and Significance

Jet stream is a strong wind blowing from west to east at high altitude, usually near the level of the tropopause (12,000 metres). The jet streams normally are 160-480 km (100-300 miles) wide by 900-2,150 metres (3,000-7,000 feet) thick, with core speed that can exceed 300 kmph (190 mph). The speed of the jet stream varies from a mean of 110 km per hour in summer to about 190 km per hour in winter. In exceptional cases a speed of 375 km per hour has also been recorded. The main types of jet stream recognised are: i) the subtropical jet stream, and ii) the polar-front jet stream of middle latitudes.

The jet stream of subtropical latitudes (30°-20°N and S of equator) is the most constant. The polar-front jet stream of mid-latitudes is less steady, owing to the transitory thermal contrasts of these latitudes.

The Arctic jet stream and the polar-night jet stream are the other jet streams observed. These jet streams are, however, weak and less steady. The velocity and location of the jet streams has had an important effect on travel times of high-flying aircrafts, especially in mid-latitudes. The polar jet stream is located at the tropopause along the polar front, at altitude between 7,600 and

10,700 metres (24,900-35,100 feet), meandering 30°N and 70°N latitudes. The polar jet stream can migrate as far south as Texas (USA), steering colder air masses into North America and influencing surface storm paths travelling eastward. In the summer, the polar jet stream exerts less influence on storms by staying far poleward.

In subtropical latitudes, near the boundary between tropical and mid-latitudes air, another jet stream flows near the tropopause. The subtropical jet stream ranges from 9,100 to 31,700 metres (29,850 to 45,000 feet) in altitude, and although it is generally weaker, it can reach greater speeds than those of polar jet stream. The subtropical jet stream meanders from 20 to 50 degrees latitudes and may occur over North America simultaneously with the polar jet stream.

The jet stream moving equatorward adopts a cyclonic curvature (anti-clockwise in the Northern Hemisphere) relative to the surface at the lower latitude as the distance from the axis of rotation increases. Contrary to this passing its original latitude, the wind takes an opposite (clockwise) curvature relative to the earth as it comes closer to the polar axis, where the rotational velocity is less. The resulting Rossby waves may have lengths of 3,000 to 6,000 km. Their length, amplitude and position are influenced by differential heating at the surface and by extensive mountain barriers. The major waves tend to persist in the westerlies above the Rockies, the Andes, and the plateaus of Central Asia and South Africa.

Winds in upper-level tropospheric waves reach maximum speeds in the jet streams, narrow bands of high velocity winds that follow the wave path near the troposphere at an elevations of 8 to 16 km. The polar front jet stream achieves its maximum force and extent in winter, when there may be two or even three distinct currents having wind speeds of 100 knots or more at their cores. Above the subtropical high pressure belts a high speed westerly flow, the subtropical jet stream, persists through most of the year. Its wind speed commonly exceeds 100 knots; maximums of more than 300 knots have been recorded in both the hemispheres, for example, over Japan and South Indian Ocean.

Near the equator, upper-level wave motion is not so well developed as at higher latitudes owing to much smaller Coriolis effect and a relatively steady flow of air that does not push vigorously across parallels of latitudes. Nevertheless, in the tropical atmosphere an easterly jet stream develops south of Asia during the summer. The tropical easterly jet is thought to be associated with thermal conditions arising from summer heating of the continent of Asia. Also known as *Krakatoa easterlies,* it probably carried volcanic dust westward after the eruption of Krakatoa in 1883.

Another atmospheric jet stream is the *polar night jet.* During winter, the sun does not heat a cone of atmosphere at the pole, and the resulting strong temperature gradient in the atmosphere creates westerly winds at heights averaging 60 km.

The global pattern of temperature, pressure and circulation migrate seasonally along the meridians. The shift in the intertropical convergence is basically the result of seasonal changes in the heat budget of different latitudes. The tendency is for motion systems to shift northward a few degrees in the Northern Hemisphere in summer and southward in winter.

Occurrence of El Nino

El Nino is the name given to the occasional development of a warm ocean current along the coast of Peru as a temporary replacement of the cold Peru current (Humboldt current) which normally operates. El Nino occurs at intervals of about three to eight years. At its occurrence a remarkable disturbance in ocean and atmosphere occurs. It begins in the east Pacific Ocean and spreads its effects widely over the globe. This disturbance lasts for more than a year, bringing droughts, heavy rainfalls, severe spells of heat and cold, or high incidence of cyclonic storms to various parts of the Pacific Ocean and its eastern coasts. The expression 'El Nino' comes from Peruvian fishermen, who refer to the *Corriente del Nino,* or the "Current of Christ Child", in describing an invasion of warm current during Christmas time which greatly depletes their catch fish. El Nino occurs at irregular intervals and with varying degrees of intensity. Notable El Nino events occurred in 1891, 1925, 1940-41, 1965, 1972-73, 1982-83, 1989-90, 1991-92, and 1994-95.

Principle of Coriolis

Most of the winds of the earth generally follow a curved path rather than a straight one. The main cause of the deflection of winds is the *rotation of the earth.*

It was established by Gaspard de Coriolis in 1844 that to an earth-bound observer, any object moving freely across the globe appears to curve slightly from its initial path. In the Northern Hemisphere, this curve is to the right (or clockwise) from the expected path; in the Southern Hemisphere, to the left (anti-clockwise). To earth-bound observers the deflection is very real – it is not caused by some mysterious force, and it is not an optical illusion or other tricks caused by the shape of the globe itself. The observed deflection is caused by the observer's moving frame of reference on the spinning earth.

The Coriolis law may be explained with the help of an illustration. Let us take an example from the equatorial city of Quito (capital of Ecuador) and Buffalo, New York. Both cities are on almost the same line of the longitude (79°W), so Buffalo is almost exactly north of Quito. Like everything else attached to the rotating earth, both cities make one trip around the world each 24 hours. Through one day, the north-south relationship of the two cities never changes – Quito is always due south to Buffalo.

A complete trip around the world is 360 degrees, so each city moves eastward at an angular rate of 15 degrees per hour (360°/ 24 hours= 15°/hour). Yet even though their angular rates are the same, the two cities move eastward at different speeds. Quito is on the equator, the 'fattest' part of the earth. Buffalo is farther north at a 'skinnier' part. Imagine both cities isolated on flat discs, and imagine the earth's sphere being made of a great number of these discs strung together on a rod connecting North Pole to South Pole. Buffalo does not have as far to go in one day as Quito because its disc is not as large. That means that Buffalo must move eastward more slowly than Quito to maintain its position due north of Quito.

The Quito disc and the Buffalo disc must turn through 15 degrees each hour (or earth would rip itself apart), but the city on the equator must move faster to the east to turn its 15 degrees each hour because its slice of the 'pie' is larger. Buffalo must move at 1,260 km (787 miles) per hour to go round the world in one day, while Quito goes at 1,658 km (1,036 miles) per hour.

The fact of being fired northward by the cannon does not change its eastward movement in the least. As the cannonball streaks north, an odd thing happens. The cannonball veers from its northward path, angling slightly to the right (east). Actually, this first cannonball is moving as an observer from space would expect, but to those of us on the ground, the cannonball 'gets ahead of the earth'. As cannonball moves north, the ground beneath it is no longer moving eastward at 1,658 km (1,036 miles) per hour. During the ball's time of flight, Buffalo (on its smaller disc) has not moved eastward enough to be where the ball will hit. If the time of flight for the cannonball is one hour, a city 398 km east of Buffalo [1,658 (Quito's speed) 1,260 (Buffalo's speed) = 398] will have an unexpected surprise. Albany may be in for some excitement.

If a cannonball were fired south from Buffalo towards Quito, the situation would be reversed. The Coriolis effect is a real effect dependent on our rotating frame of reference. Part of that frame of reference involves the direction from which you view the problem.

A quick way to remember the Coriolis effect: Northern Hemisphere/clockwise/right; Southern Hemisphere/anti-clockwise/left. The Coriolis effect also influences the path of objects moving from east to west, or west to east. Because the Coriolis effect influences any object with mass, as long as that object is moving, it plays a large role in the movements of air and water on the earth.

Atmospheric Circulation and Coriolis Effect: As stated at the outset, the air does warm, expand, and rise at equator, whereas air does cool, contract, and descend at the poles. Instead of continuing all the way from equator to pole in a continuous loop in each hemisphere, air rising from the equatorial region moves poleward and is gradually deflected eastward, i.e., turns to the right in the Northern Hemisphere and to the left in the Southern

Hemisphere. The eastward direction caused by the Coriolis effect does not produce the wind; it only influences the wind's direction.

As air rises at the equator, it loses moisture by precipitation caused by expansion and cooling. This drier air now grows more dense in the upper atmosphere as it radiates heat to space and cools. When it has travelled about a third of the way from the equator to the pole, i.e., to about 30°N-30°S latitudes, the air becomes dense enough to fall back towards the equator when it reaches the surface. In the Northern Hemisphere the Coriolis effect again deflects this surface air to the right – the air blows across the ocean and land from the northeast. Though it has been heated by compression during its descent, this air is generally colder than the surface over which it flows. The air soon warms as it moves equatorward, however, evaporating surface water and becoming humid. The warm, moist, less-dense air begins to rise as it approaches the equator and completes the circuit.

Such a large circuit of air is called an *atmospheric circulation cell.* A pair of these tropical cells exists, one on each side of the equator. They are known as *Hadley cells* in honour of George Hadley, the London lawyer and philosopher, who worked out an overall scheme of wind circulation in 1735.

A more complex pair of circulation cells operates at mid-latitudes in each hemisphere. Some of the air descending at 30 degrees latitudes turns poleward rather than equatorward. Before this air descends to the surface, it is joined by air at high altitude returning from the north. A loop of air forms between 30 degrees and about 50-60 degrees latitudes. As before, the air is driven by uneven heating and influenced by the Coriolis effect. Surface air in this circuit is again deflected to the right, this time flowing from the west to complete the circuit. The mid-latitude circulation cells of each hemisphere are named *Ferrel cells* after William Ferrel, the American who discovered their inner working in the mid-19th century.

Meanwhile, air that has grown cold over the poles begins moving towards the equator at the surface, turning to the west as it does so. At between 50 and 60 degrees latitudes in each hemisphere, this air has taken up enough heat and moisture to ascend. However, this polar air is more dense than the air in the adjacent Ferrel cell and does not mix easily with it. The unstable zone between these two cells generates most mid-latitudes weather. At high altitude the ascending air from 50 to 60 degrees latitudes turns poleward to complete a third circuit. These are *polar cells.*

Three large atmospheric cell, therefore, exist in each hemisphere. Air circulation within each cell is powered by uneven solar heating and influenced by the Coriolis effect.

Principle of Buys Ballot

The Dutch meteorologist Buys Ballot formulated a law about winds, i.e., if you stand with your back to the wind in the Northern Hemisphere, air pressure is lower on your left than on your right.

In contrast to this, in the Southern Hemisphere, again with your back to the wind, lower pressure will be on your right and higher pressure on your left.

Some of the other important laws of atmospheric circulation are:

- The winds are strong where the isobars are crowded, and weak where they are wide apart.
- The relation between isobar spacing and wind speed is rather firm in high and mid-latitudes but weakens as we approach the equator. Between 10°N and 10°S, it is difficult to relate the winds to pressure distribution.
- In the large wind systems, the air is slow starter, but when it has worked up some speed, it will carry on for a longer time.
- Along and near the earth's surface, wind does not move freely in a horizontal plain. The irregularities of the earth surface (e.g., mountains, hills, etc.) influence the direction of winds.
- Other factors being equal, the difference in wind speed and direction between the surface and upper levels is greatest over rough land surface. Over water the surface wind more nearly equals the gradient wind.
- The maximum speed of wind usually occurs in the early afternoon and the minimum in the early morning hours just before the sunrise.

- Winds are named for the direction from which they come. A wind blowing from north to south is a north wind, a wind blowing from west to east is a west wind and a wind blowing from east to west is an east wind.

Pattern of Pressure Belts

On an ideal earth (earth without a complicated pattern of land and water and no seasonal changes) there is a close relationship between the pressure distribution and wind systems.

Equatorial Low Pressure Belt: This is a zone of low pressure near and more or less parallel to the equator, where the NE and SE trade winds meet. This is also known as *doldrum.*

Doldrum is the equatorial belt of low atmospheric pressure where the NE and SE trade winds converge on, and meet each other, producing calm and light surface winds and a strong upward movement of air. The word 'doldrum' has come to be associated with a gloomy, listless mood, perhaps reflecting the sultry air and variable breeze found there. In this zone strong heating causes surface air to expand and rise. The humid, rising, expanding air loses moisture as rain, some of which contributes to the success of tropical rainforests.

Despite of the calms and light winds, the doldrums are characterised by the turbulent and stormy weather, with heavy rains, thunderstorms, and squalls. On account of there lack of wind, they used to be avoided, as far as possible, by sailing ships for the latter were often becalmed for several days.

The doldrums are variable both in position and extent, usually moving northward and southward with the sun; their movement, however, is much less than that of the sun, being about 5 degrees on each side of the mean position of equator, with a lag of one or two months behind the sun.

Subtropical High Pressure Belts: The subtropical high pressure belts lie adjacent to but just outside the Tropic of Cancer and the Tropic of Capricorn. The subtropical high pressure belts extend upto 40°N and 40°S. These belts are the main regions of anticyclones. The subtropic highs are, however, not continuous belts but are broken up into 'cells' having their best development over the oceans.

In the subtropical belts the air generally descends. This descending air is generally arid. The great hot deserts of both the hemispheres, dry bands centred round 30 degrees, mark the intersection of the Hadley and Ferrel cells. Because evaporation is higher than precipitation in these areas, ocean surface salinity tends to be highest at these latitudes. At sea, the areas of high atmospheric pressure and little surface wind are called the *horse latitudes.* The Spanish ships laden with supplies for America and West Indies were often becalmed there, sometimes for weeks on end. When the mariners ran out of water and feed for their livestock, they were forced to throw the roped horses into the sea. The ships passage used to be unduly delayed.

Subpolar Low Pressure Belts: The subpolar low pressure belts lie between 60 and 65 degrees latitudes in both the hemispheres. Owing to low temperatures in these latitudes the subpolar low pressure belts are not very pronounced. These are dynamically produced by the rotation of the earth on its axis. In fact, the surface air spreads outward from this zone due to the rotation of the earth which results into low pressure belt on the subpolar areas. The subpolar low pressure belt is more developed in the Southern Hemisphere than in the Northern Hemisphere. The North Pacific *Aleutian Low* develops near Aleutian Islands, and the North Atlantic *Icelandic Low* develops near the Iceland. Both these cells are dominant in winter and weaken significantly or disappear altogether in the summer with the strengthening of high pressure in the subtropics. The area of contrast between cold and warm air masses forms a contract zone known as the *polar front* which encircles the earth at 60 degrees latitudes and is focused in these low pressure areas.

In the Southern Hemisphere, a non-continuous belt of subpolar cyclone systems surrounds Antarctica.

Polar High Pressure Belts: The Arctic and Antarctic polar regions are cold throughout the year. These are the areas of high pressure characterised by permanent anticyclone. Easterly winds spiral outward from the polar high pressure belts. Of the two polar regions, the Antarctic high is significant in terms of both strength and persistence. In the Arctic, a polar high-pressure cell is less pronounced, and when it does form, it tends to locate over the colder northern areas in winter (Canadian and Siberian highs) rather than directly over the relatively warmer Arctic Ocean.

Four Hemispheric Pressure Areas

Name	*Cause*	*Location*	*Air Temperature/Moisture*
1. Polar high pressure belt	Thermal	90°N, 90°S	Cold/dry
2. Subpolar low pressure belt	Dynamic	60°N, 60°S	Cool/wet
3. Subtropical high pressure belt	Dynamic	20° to 35°N and S	Hot/dry
4. Subtropical low pressure trough	Thermal	10°N to 10°S	Warm/wet

Winds and their Role

The general circulation of the atmosphere is known as the *permanent wind system.* The permanent wind (planetary wind) system consists of: i) the trade winds, ii) the anti-trade winds, and iii) the polar winds. These permanent winds are produced by global pressure distribution and pressure gradient. There are, however, some local winds, also known as *named winds,* which blow in narrow areas (localities) and influence the human health and their socio-economic activities.

The local winds (regular or periodic) are appreciably distinctive as a result of their characteristics of temperature and humidity. Thus, the local winds in general are developed as a result of local temperature, pressure and humidity variations. Their origin is also attributed to the formation of air currents, crossing mountain ranges, valleys and physical barriers. The main causes for the development of local winds are:

1. Unequal heating of land and sea resulting into the land and sea breezes.
2. Heating and cooling of the mountain slopes.
3. Local winds originating because of the deformation of air currents, crossing of mountain ranges, and physical barriers.
4. Convectional local winds, caused by steep pressure gradients and steep variations in local temperatures.

Thermal Theory

The thermal concept about the origin of monsoon was advocated for the first time by Halley in 1686. Accordingly, there is unequal distribution of land and water in the world and these two surfaces (land and sea) react differently to the incoming solar radiation. The land heats and cools quickly while the ocean water, being liquid, becomes warm and cool gradually. In other words, the conductive capacity of the oceans is very much larger than that of the landmasses. For this reason, the annual variation of temperature is very much smaller over oceans than continents. Furthermore, the annual variation of the incoming radiation is small near the equator and increases towards the poles. As a result, the annual variation of temperature will be small in low latitudes and large in high latitudes. It follows that the annual variation in the temperature difference between oceans and continents will be small near the equator and large in high latitudes.

In spring (April), when the land warms much faster than the oceans, an area of low pressure develops over it while the pressure rises over sea. The process is just like that of the sea breeze. However, in the annual cycle, the changes are so slow and the system so large that the winds are very nearly geostrophic. Moreover, the annual variation penetrates to great heights. Thus, in summer, there is a tendency for low pressure to develop over the continents. These are warm lows, and their intensity tends to decrease with elevation.

In autumn (October and November), when the land cools, a high pressure area develops over land, and the pressure lowers over sea. The temperature difference between land and sea is much larger in winter than in summer, with the result that the continental high in winter is far more pronounced than the continental low in summer. Furthermore, the continental high in winter is distinctly cold, and its intensity decreases rapidly with elevation.

The Indian monsoon is the best known example of an alternating circulation system which develops in response to the annual variation in the temperature difference between oceans and continents. Since South Asia is situated close to the equator, the monsoon is considerably strengthened by the annual migration of air from the winter to the summer hemisphere, and this migration is somewhat intensified by the annual heating and cooling of Australia.

During the winter solstice (winter in the Northern Hemisphere) when the sun becomes vertical over the Tropic of Capricorn in the Southern Hemisphere, high pressure area develops near Baikal Lake and Islamabad (Pakistan). Contrary to this, low pressure area develops over the southern Indian Ocean. Consequently, winds start blowing from the high pressure land areas to the low pressure oceanic areas. These winds are called as the *winter* or *northeast monsoon*. The winter monsoons are generally dry as they originate from the land areas.

In the summer season, the above pressure conditions get reversed. At the time of summer solstice, the sun's rays fall vertical at the Tropic of Cancer, leading to high temperatures and low pressure in the Northern Hemisphere. In the summer season, in the Northern Hemisphere, two low pressure areas develop, one near the Baikal Lake and the second near Multan (Pakistan). Contrary to this, the high pressure area develops over the Indian Ocean and the Pacific Ocean (south of Japan). Consequently, wind blows from the Indian Ocean to the landmass of Asiatic continent. These winds in Indian Ocean, while crossing the equator, become southwesterly (Ferrel's law). On their way, while passing over the water of the ocean, these winds pick up moisture and result into rainfall when obstructed effectively by the mountain barriers like the Himalayas and the Western Ghats. These are the rainfall giving southwest or the summer season monsoon. The summer monsoon reaches a maximum of intensity in the month of July.

The summer monsoon begins in April in Myanmar (Burma) and in May in Kerala coast; it is however, retarded in the Gangetic plain of India. During a three-week period in late May and early June, the monsoon sweeps northward into central India. It is accompanied by heavy rains, squalls, and occasional violent storms of the type known as *typhoons* (toofans) in China, and *hurricanes* in the United States. On its western side the summer monsoon is limited rather sharply by warm, dry air of great depth which comes from the deserts to the west and northwest. This accounts for the sharp transition from the Thar desert to the neighbouring regions with broad current over South East Asia, bringing much rainfall with it.

The thermal concept about the origin of monsoon has not been accepted universally and been criticised on several counts as under:

1. The low pressure areas that develop over the continents are not stationary. These low pressure areas change suddenly which reveal that the low pressure areas are not exclusively related to low thermal conditions. In fact, they represent the cyclonic laws associated with the southwest monsoon.
2. The rainfall associated with the monsoon is the result mainly of the tropical cyclones.
3. Had the monsoon been thermally induced, there should be anti-monsoon circulation in the air, but some of the meteorologists have observed the occurrence of low pressure areas in the upper atmosphere (near the tropopause) also. Such upper air winds, which change their directions seasonally, are often called the *upper air monsoon* or *aerological monsoon.*

Flohn's Theory

Looking at the inadequacies of the thermal concept, Flohn propounded the dynamic concept about the origin of the monsoon. According to this, monsoons are originated due to shifting of pressure and wind belts. Tropical convergence is formed due to convergence of northeast trade winds near the equator. This is called as the *Inter-Tropical Convergence* (ITC). The northern and southern boundaries of the ITC are called *NITC* and *SITC*, respectively. There is a belt of doldrum within the ITC characterised by *equatorial westerlies.* At the time of summer solstice (June 21), when the sun's rays are vertical over the Tropic of Cancer, NITC is extended upto 30°N latitudes covering south and South East Asia and thus equatorial westerlies are established over these areas. These equatorial westerlies become southwest or summer monsoon. The NITC is associated with numerous atmospheric storms (cyclones) which result into heavy rainfall during the season of general rains or *barsat* (mid-June to mid-October). The northeast or winter monsoon does not originate due to low pressure in the Southern Hemisphere during the winter solstice (23 December) when the sun's rays fall vertical over the Tropic of Capricorn. The ITC, consequently, pushes south of equator. The northeast trade winds blow from the land towards the sea/oceans and thus become winter monsoon. Since they originate from the land, they are dry and give rainfall only in those parts where they pass over the sea surface. The Eastern Ghats as well as eastern parts of Sri Lanka receive rainfall from the northeast monsoon.

Origin of Indian Monsoon

The monsoon wind systems are best developed in India and the adjacent countries of South East Asia. The mystery of monsoon and its erratic behaviour have been explored by the meteorologists for the last five decades. The data available from the meteorological stations all over the world and the imageries obtained from the satellites show that the Indian monsoon is largely controlled by the following factors:

1. The differential heating of the landmass of Asia and the Indian Ocean.
2. The existence of the Himalayan ranges and the Tibetan plateau.
3. The occurrence of heavy-light snow over the Tibetan plateau.
4. The existence and circulation of upper air jet streams in the troposphere.

The origin of Indian monsoon has traditionally been ascribed to the heating and cooling of the continent of Asia with the changes of season and resulting changes in the distribution of pressure. This simplistic model does not explain many of the complex issues associated with monsoon. The recent studies about the upper air circulation of atmosphere show that the monsoon is a highly complex phenomenon and is also closely influenced by the middle and upper tropospheric movements.

The occurrence and changing position of jet stream influence the origin and development of monsoons. In winter, the high pressure over the continent of Asia forces the westerly jet stream southward so that a portion of it lies south of the Himalayas.

Subsiding air along the tropical margins of the jet moves towards the Indian Ocean, blocking incursions of maritime tropical air and producing the dry winter monsoon. In summer, the jet stream shifts with the general circulation and lies north of the Himalayas, allowing a series of low-pressure disturbances to bring moist tropical air of the wet summer monsoon onto the continent.

The presence of the Himalayan ranges and Tibetan plateau also have a close bearing on the origin of monsoon. The jet streams that lie about 12 km in the troposphere are bifurcated by the Himalayan mountains and the Tibetan plateau. The northern branch blows from west to east in arcuate shape to the north of the Himalayas and Tibetan plateau, while the southern branch moves from west to east to the south of Himalayas. It is interesting to note that the main branch of upper air jet streams follows anticyclonic path wherein anticyclonic air circulation is developed to the right of the general flow direction of the jet streams across Afghanistan and Pakistan, with the result upper air high pressure is formed over them. Contrary to this, the main branch of the jet streams to the south of the Himalayas follow cyclonic arc, having anti-clockwise air circulation due to the mountain barrier. Consequently, the upper air low pressure and cyclonic air circulation are developed over Tibetan plateau. Thus, the Indian monsoon is considerably affected by the interrupting effect of the lofty and extensive plateau of Tibet on the upper westerlies. In summer, these shift north of the plateau, emphasised by the northward migration of the subtropical jet, thus permitting the northward surge of the southwest monsoon.

The monsoon is quite conspicuous in the subcontinent of India, mainly because of its neighbourhood to equator, and Indian Ocean in the south and high and vast transverse relief barrier (Himalayas) in the north. The rainfall distribution in the subcontinent is highly unequal. The main cause of spatial variation in rainfall is relief features. For example, the Arabian Sea branch of the southwest monsoon ascends abruptly after being obstructed by the Western Ghats which gives heavy rainfall to the windward side. Contrary to this, on the leeward side of the Western Ghats, the monsoon winds descend and are thus warmed due to which the relative humidity decreases. Consequently, the leeward side of the Western

Ghats receives very little rainfall. Similarly, the Himalayan mountain ranges affect the Bay of Bengal stream of the southwest monsoon. Owing to the presence of the Himalayas, the air ascends, and gives heavy rainfall to the windward (southern) slope, while the leeward slope (northern) of the Himalayas remains almost rainless. The Himalayan ranges also help in the westward movement of the summer monsoon. The areas which lie to the west of the Aravallis receive very scanty rainfall.

Monex Expedition

Monex is a joint expedition of India and Russia which started working in 1973. On the basis of data generated and processed by Monex, the meteorologists opine that the Indian monsoon is related to the heating and cooling of the Tibetan plateau (having widths of 1,000 km and 600 km in the east and the west, respectively, and length of 2,000 km and height about 4,000 m). The influence of the Tibetan plateau on the genesis of monsoon has already been discussed in the preceding paras. The summer monsoon has a very close influence on the life of the subcontinent of India. The Indian agriculture and its allied activities are largely governed by the timely arrival and amount of rainfall. It also controls the cropping patterns, yield, and production of the cereal and cash crops. The periods of marriages, festivals, and cultural ethos are also adjusted according the arrival and retreat of the summer monsoon. Despite irrigation development, the Indian agriculture is still known as a *gamble on monsoon.*

Behaviour of Monsoon

The erratic behaviour of monsoon and delays in the onset or the *'burst'* of the summer (wet) monsoon have disastrous consequences for agriculture, industries, natural vegetation and water supplies in the monsoon region, especially in South Asia. In Shillong (Meghalaya), the non-occurrence of rainfall even for one week may be disastrous. At such occasions, in Shillong, people purchase buckets of water at expensive rates. For socio-economic and scientific reasons, meteorologists continue to seek the physical causes of monsoon and attempt their prediction. One of the probable causes of a delayed summer monsoon over northern India is heavy winter snow on the Tibetan plateau, where the resulting increase in surface albedo retards spring warming and normal development of the required circulation pattern.

The monsoon climate may be regarded as transitional in nature from the rainy tropics to the wet and dry tropics, having annual rainfall totals comfortable to the former and a regime of precipitation similar to the latter. It is well to remember that the term *'monsoon tropic'* does not apply to all climates affected by a monsoonal wind circulation. Its use stems from the characteristic climates of Monsoon Asia, but the designation wet-and-dry tropics is applied to those regions with less annual precipitation and a distinct season of drought.

Mean monthly temperatures in the monsoon tropics are not greatly different from those in the rainy tropics — typical values ranging well above 20°C. These climates extend farther poleward and the annual range is greater in some cases. The maximum temperatures usually occur in the months of April and May, the period of clearer skies, increasing insolation, and water budget deficit just before the onset of more persistent cloudiness and heavy rainfall. Kolkata is a representative station of the typical monsoonic climate.

Kolkata: Mean Monthly Temperature and Rainfall

	J	*F*	*M*	*A*	*M*	*J*	*J*	*A*	*S*	*O*	*N*	*D*	*Yr*
T(°C)	20	22	27	30	30.5	30	29.5	28.5	29	27	24	20	26.5
Rainfall in cm	1.0	3.0	3.5	4.5	14.0	29.0	32.0	33.0	25.5	12.0	2.0	0.5	160.0

The diurnal variation of temperature in the monsoon tropics is, on the average, slightly greater than in the rainy tropics. It is greatest in the drier months and least in the rainy season. During the winter season, there may be some influence from cyclonic disturbances which pass on the poleward margin with resulting short periods of comparatively low temperatures.

The annual rainfall averages above 1,500 mm in most areas of the monsoon tropics. Where a strong onshore flow of moist air meets a coast backed by a mountain barrier, exceptional rainfall totals are achieved.

Chennai: Mean Monthly Temperature and Rainfall (13°N, 82.5°E, height 16 metres above the sea level)

	J	*F*	*M*	*A*	*M*	*J*	*J*	*A*	*S*	*O*	*N*	*D*	*Yr*
Temperature (°C)	24	26	28	30	33	32	31	30	30	28	26	25	29
Rainfall (mm)	24	07	15	25	52	53	83	124	118	267	308	157	1233

Rainfall which accounts for most of the total in the monsoon tropics is of the showery type, with the orographic effect often playing an important part, but certain areas are visited by tropical depressions. Summer weather, i.e., the rainy season, is similar to that of the rainy tropics, usually with greater monthly rainfall totals. Winter is slightly cooler and somewhat drier, having protracted sunny periods with occasional thunder showers. Remnants of polar fronts sometimes invade these climates above the level of the trades in winter, producing overcast skies and light rainfall but no very great temperature decrease at the surface.

The monsoonal effect is felt most markedly in Asia because of its great size, and also because of its vast transverse mountain barrier. To a lesser extent, North Australia, North America, coasts of western South America, West Africa, Malagasy, Madagascar, etc., experience the monsoons characterised by the reversal of winds.

The Basis

The meteorologists and climatologists are actively probing the causes of origin of monsoon. The data and information made available by the satellites have solved many of the complex problems about the origin of monsoon. The experts of meteorology are, however, not unanimous about the causes of origin, development and erratic behaviour of monsoon. At present, however, there are two concepts in vogue about the origin of monsoon. These are: i) the thermal concept, and ii) the dynamic concept.

Different Local Winds

The number of local winds is enormous. Some of them, however, are of great meteorological interest as they directly influence the health, efficiency and economic activities of the people of the region concerned. The local winds also influence the standing crops, orchards, vegetables, flowers and fruits.

Land and Sea Breezes: Land and sea breezes are local periodic winds on a diurnal rather than on a seasonal basis. They result when differential heating takes place within a relatively short distance — a condition which occurs most commonly near the sea coast. Their air movements may be generated either by heating or cooling of a particular area.

The *sea breeze* develops along sea coasts or large inland water bodies (lakes) in summer when the land heats much faster than the water on a clear day, and a pressure gradient is directed from high over the water to the low over the land. Under these conditions, in the day, the air blows from the sea to the land. The air movement from the sea to land reaches the greatest intensity during the afternoon. In the equatorial region, it is a regular phenomenon, where it brings some amelioration to the hot humid coast lands. Elsewhere it is generally a summer phenomenon. It varies considerably in strength along the same coast, being influenced by topography.

The sea breeze rarely has a depth of more than 1,000 metres and its maximum inland strength does not extend more than 50 km. It begins offshore in the late morning hours and gradually extends inland to decrease afternoon temperatures. Towards evening it subsides. Because of sea breeze, areas on the immediate coast in middle latitudes and the tropics may have lower temperatures than a few kilometres inland. The sea breeze brings cool air off the water, dropping temperatures, and refreshing beach-goers and residents living close to the beach.

The *land breeze* is a diurnal wind blowing from the land out to sea. It is caused by the differential cooling of land and sea. The cooling of the air over the land by radiation during the night causes the air to descend and flow seawards. It is confined to coastal regions and lake sides, and, though mostly characteristic of tropical lands, may also be observed in the temperate zone during summer. It is most developed when the general pressure gradient is slight and the sky clear; it may then set in about midnight or a few hours later. The land breeze is much influenced by topography, and varies considerably in different parts of the coast.

In general, the land breeze is less extensive both vertically and horizontally than the sea breeze. It usually attains its maximum intensity in the morning hours and dies out soon after sunrise.

Both the land and sea breezes are most marked in normally calm areas, for on the coasts of trade wind islands, for example, their effect is marked by the dominant winds. On the islands and along the coasts near the equator, however, the sea breeze brings a welcome relief from the stagnant humid heat. In some islands these winds are so regular that fishing boats go out at night with the land breeze and return the next afternoon with the sea breeze.

In the Northern Hemisphere, the mature sea breeze blows with the land on its forward left side, and in the Southern Hemisphere the land is on the forward right side. Sea and land breezes are examples of how heating and cooling produce a pressure gradient that causes a wind.

Mountain and Valley Winds: Mountain and valley winds may be termed as periodic winds which undergo a complete daily reversal. They alternate in direction in a manner similar to the land and sea breezes. On mountain sides, under a clear night sky, the higher land radiates heat faster and is cooled faster. The cool, denser air then flows down the mountain slopes into the valleys and lowlands. Since it blows from the mountain, this air flow is termed as *mountain breeze.* By morning it may produce temperature inversion in depression, so that the valley bottoms become colder than the hill sides. On warm sunny days, the heating of mountain slopes may generate an upslope flow of air called a *valley breeze.* As the warm air moves up the mountain, it is replaced by cooler air from above the valley, and surface temperatures are moderated slightly. In general, the mountain winds (night winds) are stronger than the valley winds (day winds).

The mountain and valley winds may be overshadowed by a general wind system. On an otherwise calm day, the mountain and valley winds reveal their presence by cumulus clouds forming over the mountains during the day and dissolving, rather disappearing in the evening.

A very pronounced form of mountain wind is where cold air blows off an ice cap, over which chilling is intense. This can be felt in Greenland, Arctic Islands, Siberia, Norway, Sweden, Alaska, Ecuador, Andes, Rockies and Himalayan mountain ranges. In Andes mountains (Ecuador), these winds are known as *nevados*.

Katabatic (Stroph or Mountain) and Anabatic (Valley) Winds: Katabatic is a local wind caused (often at night) by the flow of air, cooled by radiation, down mountains slopes and valleys. It is also caused by the flow of cold air down the slopes of ice caps, such as Antarctica and Greenland. It is caused by ground surface cooling as a result of radiation, which in turn cools the lower air layers. With the rapid loss of heat by radiation, the mountain or ice-cap becomes cold, and the chilled air moves downward under the action of gravity. The direction of flow is controlled almost entirely by orographic features.

Anabatic is an upslope wind formed when air on hill sides is heated by insolation conduction to a greater extent than air at the same horizontal level but vertically above the valley floor. This causes convectional rising of the heated air, which is replaced by cooler air from the valley floor.

Drainage (Gravitational) Winds: During the cold season, in the temperate latitudes, a high pressure area develops over plateaus and inland areas sheltered by mountains. Some of the cold air seep down the slopes, gather in the valleys and fiords, and come to the coast as a gentle or moderate breeze. However, when a travelling disturbance (cyclone) approaches, the cold air accelerates out through the mountain gaps, valleys and canyons, and arrives as a cascade of cold air with strong and gusty winds. Although the air warms up adiabatically while it descends, the temperature difference between inland and coast is normally so large that the air arrives as a cold current. These winds are particularly strong, and sometimes destructive, where a large reservoir of cold air has to empty itself through a narrow gap or valley, or where several converging valleys meet. A **wind** of this type is generally called as *katabatic,* since it is caused by gravitation of cold air off high ground. Because the air is steered between the walls of a valley, the drainage winds show little relation to the isobars. Often, the wind blows right across the isobars from high to low pressure.

Drainage winds occur, with variable strength, in all mountainous regions, particularly where mountains are close to a coast. In Greenland, Spitzbergen and the Antarctic, such winds are frequent and sometimes destructive. The drainage winds may be classified under: a) depression winds, and b) descending winds.

Gloomy Winds

Some of the important depression hot winds are: Brickfielder, Chili, Gibli, Harmattan, Karaburn, Khamsin, Leveche, Loo, Sirocco and Zonda. There are also numerous depression cold winds, of which the Blizzard, Bora, Buran, Friagem (Surazo), Gregale, Levanter, Maestro, Mistral, Nevados, Norte, Norther, Pampero, Papagayo, Polar Outbreak, Purga, Solano, Southerly Burster, Tramontana, Vendavales and Williwaw are worth mentioning.

Brick Fielder: It is a hot, dry, northerly wind which blows from the Australian interior out to the southeast coastlands during the summer (December, January and February), and often brings clouds of dust in the form of dust storms. The wind has a temperature of 38°C/100°F. It blows with greater force in Victoria and the southwestern parts of New South Wales. It moves ahead of a depression or a trough of low pressure, and precedes the southerly burster.

Chili: It is a hot, dry, southerly wind of Tunisia (North Africa). It blows from the Sahara desert of Algeria and moving northward through Tunisia, enters the Mediterranean Sea. Its temperature may be as high as 40°C and the relative humidity falls to less than 10 per cent.

Gibli: It is a hot, dry, southerly wind of Libya (North Africa). It generally lasts from two to three days. Its temperature varies between 38°C and 45°C. At its occurrence, the outdoor activities get suspended, the people remain indoor and often the government declares holidays. When blowing with a force of gale, the dried grasses catch fire and cause damage to orchards, people and property.

Harmattan: It is a strong northeasterly wind experienced in West Africa. Blowing direct from the Sahara desert, it is so hot, dry and dusty that sometimes splits the trunks of trees. Through Mali and Niger, when it penetrates to the Guinea coast in winters, it provides a welcome relief from the moist heat, and is beneficial to health. Consequently, it is sometimes known as *the doctor.* On the other hand, it carries with it abundant dust from the desert, and a thick haze is sometimes formed which impedes river navigation. It may cause severe damage to crops. It is so dry and dusty as to be often injurious to health. Its average southern limit in winter, when the equatorial belt of low pressure is along or just south of the Guinea Coast, is about 5°N; in mid-summer, however, the southern limit is about 18°N.

Khamsin: It is a hot and dry southerly wind experienced in Egypt. In character, it resembles to the Sirocco of North Africa. According to the Arabs, it blows for a period of 50 days, from April to June, the name 'khamsin' being the Arabic word for fifty. It blows ahead of the depression which move eastward along the Mediterranean Sea or across Africa, and often carries with it considerable quantity of dust from the interior south. In the Middle East, the name is also rather loosely applied to any hot, dry wind blowing off the desert.

Leveche: It is a hot, dry, southerly wind, originating from the Sahara (Algeria) and experienced in southeast Spain, also corresponding to the Sirocco of North Africa. It blows ahead of an advancing depression and often carries a great deal of dust.

Loo: It is a hot and dry wind which blows in the summer months of April, May and June in Rajasthan, Gujarat, Haryana, Punjab, Uttar Pradesh and western and northern parts of Madhya Pradesh. Its strength is subject to considerable diurnal variations. Its movement is usually feeble during the night hours and varies very little from 7 pm to 8 am. It then increases steadily upto about 1 pm or 2 pm. Occasionally, when conditions are most favourable, it blows with the force of gale during the next two or three hours, after which it falls off again very rapidly until 7 pm, when the wind is nearly calm at night.

It blows from west and southwest direction and there is little shift in its direction during the day hours. The origin of this wind is attributed to the convective air movement, produced by the heating of the surface soil and the rapid decrease of temperature as one goes up in the lower strata.

Sirocco (Scirocco): It is a very hot, dust-laden, south or southeasterly wind which blows from the desert of Sahara towards the Mediterranean Sea, especially during the summer season (April to July). Although it starts as a dry wind, it picks up moisture as it crosses the North African coast and by the time it reaches Malta, Sicily and Italy, it becomes a humid, oppressive wind causing discomfort. It is formed ahead of one of the depressions which pass from west to east along the Mediterranean Sea; it may therefore occur in all seasons, but it is very rare in summer, and is specially important in spring, when the depressions are most vigorous and the desert is already hot. It usually lasts for a day or two, and is then replaced by a much cooler northerly wind behind the depression. It withers vegetation and often causes much damage to crops, especially if it blows while the vines and olives are in blossom. In Egypt, the wind is known as the *khamsin,* in southeastern Spain as the *leveche,* in Tunis as the *chili,* and in Libya as the *gibli.*

Zonda: In Argentina and Uruguay, a hot sultry wind bringing tropical air from the north is known as *zonda.* It is an analogous to the Australian brickfielder, and usually preceding the pampero. It has a very enervating effect on the inhabitants. The name is also used for a warm, dry wind of the 'foehn' type which blows from the Andes, similar to the Chinook of North America.

Blizzard: An intensely cold, strong, and abnormally high wind accompanied by falling snow. At the occurrence of a blizzard the visibility is practically reduced to zero. Some of the snow falls from the clouds, but much of it is swept up from the ground. In Canada and the northern United States, the blizzard, a special type of cold wave, often disturbs the calm of the winter anticyclone, coming with a northerly wind of gale strength in rear of an eastward-moving depression. It is most dangerous over the open prairies. In polar regions, blizzards are frequent in some localities, at times excessively severe, and may continue for some days.

Terre Adelie in Antarctica, for instance, is especially subject to them, and has been called *the home of the blizzard*. There a sudden rise of temperature occurs at the commencement of a blizzard in winter, as the surface temperature inversion is destroyed; in summer the temperature falls with a blizzard.

Bora: A strong, cold and often very dry northerly or northeasterly wind experienced along the eastern coast of the Adriatic Sea and in the northern Italy, mainly in winter (December to March). It occurs when atmospheric pressure is high over Central Europe and Balkans (Bulgaria, etc.) and low over the Mediterranean Sea. The mountainous topography helps to funnel the wind, which occasionally reaches a speed of 135 kmph (80 mph) with gusts upto 225 kmph (135 mph).

Blowing from the high pressure area of the Balkan mountains, it usually gives clear skies and cold, dry weather, but if it is associated with a depression over the Adriatic Sea, it may be accompanied by heavy clouds and rain or snow. If the continental anticyclone is well established, it may continue for several days. It often blows with great strength, with violent gusts and squalls.

Friagem (Surc zo): The cold wave experienced on the tropical campos of Brazil during winter, caused by the development of an anticyclone. In the region of the middle Amazon, for instance, there is sometimes a cool spell during May or June lasting several days. The temperature may fall below 10°C, causing acute discomfort, and rendering the natives extremely vulnerable to colds.

Gregale: A strong wind from the northeast direction in the north-central Mediterranean area of the Adriatic coast, blowing mainly in the cool season. It occurs most frequently when pressure is high to the north over Central Europe and the Balkans, and relatively low to the south over Libya. This wind brings little or no rain and the temperature does not fall much below the normal. Similar names (e.g., gregal in southern France) have been applied to strong winds from a similar direction in other Mediterranean areas.

Buran: It is a strong, cold northeasterly wind experienced in Siberia and Central Asia. It blows mainly during the winter

(December to March) in rear of depression, breaking the comparative calm of anticyclonic conditions, carrying with it snow and ice particles. It thus corresponds to the Canadian blizzard. It reaches a gale force, often blows at temperatures of -30°C (-20°F) or below, and is thus dangerous to both human and animal life, especially in the open steppes, as the blizzard is on the prairies. The winter 'buran' is also termed as the 'purga'.

Levanter (Solano): It is a local wind which blows in the western parts of the Mediterranean Sea. Its main path lies between the Balearic Island and the Spanish mainland. Blowing westward it passes through the Strait of Gibraltar. This occasional wind blows in the summer and early autumn, when a depression enters the western Mediterranean Sea. It is associated with storminess and heavy rains. It causes dangerous eddies on the lee side of the Rock of Gibraltar *(Gebl-e-Tariq)*. With a moderate wind strength, a *banner cloud* extends from the summit of the Rock of Gibraltar a kilometre or more to leeward; when the wind has gale force, the cloud lifts and disappears. The wind is also known as *solano*.

Maestro: It is a northwesterly wind experienced in the central Mediterranean area, most strongly on the western side of a depression, especially in the Adriatic Sea, the Ionian Sea, and round the coasts of Corsica and Sardinia.

Mistral: Mistral is a cold northerly or northwesterly wind experienced on the shores of northwest Mediterranean Sea, especially in the Rhone Valley (France) and its delta. It is most prevalent during the winter, when pressure is relatively high over the western Mediterranean. It reaches gale force in the rear of a deep depression situated over the Gulf of Lions which is moving away eastward or southward. The air sweeps southward from the Central Plateau of France, is funnelled through the Rhone Valley, and thus reaches the delta as an extremely strong wind, as well as being cold and dry. The wind strength on the surface is often 60 kmph, but speed over 130 kmph has been recorded, and trains have been overturned. It is of special significance to aviation as it was in the past to navigation ships. When the wind blows, the sky is often cloudless, but the wind frequently brings temperatures well below freezing point. Along the lower reaches of the Rhone, its power is displayed by the permanent set of the trees towards the southeast, orchards and gardens are protected from it by thick hedges of cypress, and many of the smaller houses have doors and windows only on the southeastern side to avert its adverse influence.

Nevados: It is a cold wind which blows down from the mountains with considerable regularity in the higher valleys of Ecuador. It is caused by the cooling of the air on the mountains, partly by nocturnal radiation, and partly by contact with ice and snow, the cold air flowing down the slopes.

Norte: It is a northerly wind experienced in winter in Central America. It is a continuation of United States' *norther,* and likewise brings a sudden and often considerable fall of temperature. At its occurrence the temperature may be reduced by 5°C to 10°C for a day or two. It is a severe change for such latitudes and the wind may blow with gale force on exposed coasts. On the northern coasts of the Isthmus of Tehuantepec, where the wind is strong and almost constant throughout the winter (December to February), it is accompanied by heavy rainfall, and when it reaches the Pacific side, on the Gulf of Tehuantepec, it is still cold but dry. The name is also applied to the cold, northerly wind experienced in eastern Spain during winter, when the high atmospheric pressure of the interior produces an outstanding current, and to the warm humid, northerly wind associated with a depression in Argentina.

Norther: The cold wave sometimes experienced in the southern United States, usually after a depression (cyclone). At the occurrence, the temperature may fall by about 20°C in 24 hours, causing great damage to fruit crops. It often blows with great violence, possibly 60 to 100 kmph. It is generally accompanied by severe thunderstorms and hails, and may give line-squall (stormy weather) conditions. In the Sacramento Valley of California, however, this wind is often dry and dusty, having been heated and dried by descent from the mountain.

Pampero: It is a violent burst of cold, polar air experienced on the *pampas* of Argentina and Uraguay after a depression (cyclone) has passed. It blows from a southerly to westerly direction, and takes the form of a line-squall, with the typical roll of cloud.

It is sometimes accompanied by rain, thundering and lightning, and dust is swept up from the pampas. A considerable fall of temperature occurs as the storm passes. It is most frequent in summer (December to March), and can be compared with the southerly burster of Australia.

Papagayo: It is a cold, northerly local wind occasionally experienced on the Mexican plateau. It is a continuation of the *'norther'* of the United States, and comparable with the 'norte' of the coastal region.

Polar Occurrence

The occasional incursion of a cold air mass, bringing strong winds and cool weather, from middle latitudes to lower latitudes. The leading edge of a polar outbreak is a cold front with squalls, which is followed by usually cool, clear weather with strong, steady winds. It is best developed in the Americas. Outbreaks that move southward from the United States into the Caribbean Sea and Central America are called *northerns* or *nortes* while those that move from Patagonia into tropical South America are called *pamperos*. A severe polar outbreak may bring sub-freezing temperatures to the highlands of South America and severely damage such cash crops as coffee.

Purga: It is a cold, northeasterly wind of Siberia. It is characterised with snow, especially so called in the Tundra. It resembles to the 'buran' wind of Siberia and Central Asia. At its occurrence, the temperature falls upto -35°C, which is hazardous to both human and animal life.

Solano: It is an easterly wind which blows in the western parts of the Mediterranean Sea along the coast of Spain. It mainly occurs during the months of summer (June-July) and early autumn (August-September) and brings moderate to heavy rains in southeastern Spain and along the Strait of Gibraltar. It is also known as *levanter*.

Southerly Burster: It is a cold polar wind, originating from Antarctica, moving northward and experienced in southern and southeastern Australia (South Wales, Victoria), usually behind a trough of low pressure. It is accompanied by typical line-squall conditions, often being heralded by a long roll of cloud. The wind changes suddenly from a warm northerly, the brickfielder, to violent cold southerly, often laden with dust and generally accompanied by thunderstorm. The fall of temperature is sudden and considerable, usually amounting 10°C to 15°C. It is associated with veering gale force winds, a thick arch of cloud, hail and thunderstorms, a rapid rise of pressure and fall in temperature. The southerly burster is most frequent in spring (October-November) and summer (December-February), and is felt most severely along the coast of New South Wales; the mountains probably impede the advance of the following anticyclone, a steep pressure gradient is created, and the southerly wind blows with extreme violence. It corresponds to the pampero of Argentina (South America).

Tramontana: In the central Italy, along the Mediterranean coast, a cool, dry northerly wind, originating from beyond the Alps mountains, is known as *tramontana.*

Vendavales: It is a strong, squally, southwesterly wind experienced in the Strait of Gibraltar and off the east coast of Spain. Being associated with depression, it blows chiefly in the winter season and brings considerable rains.

Williwaw: It is a violent squall experienced in the Strait of Magellan where the winds are almost constantly strong and westerly.

Winds, which are Descending

Among the descending hot winds, Berg, Fohn, Chinook, Norwester, Samoon, Santa Anna, Aandhi, Dust Devil, Haboob, Karaburn, Shamal, Siestan and Simoom are important.

Berg: 'Berg' is a term of German and Dutch derivation, used locally to describe a single hill or range of mountains. The local wind that has its origin in a hill or mountain range in Germany is known as *berg wind.* It is a hot, dry wind that blows down to the central part of Germany. It often leads to oppressive weather. The combination of heat and excessive dryness, together with strong, gusty winds, causes physiological as well psychological reactions, irritableness, headache, etc. In South Africa similar winds blow from dry plateau down to the sea coast.

Fohn (Foehn, Fon): It is a warm, dry wind which blows down the leeward slope of a mountain, best known in the valleys of the northern Alps, where the name originated. It occurs when a depression moving to the north of the Alps draws in air from the south. This air ascends the southern slopes, clouds are formed, and there is heavy rain; when it reaches the northern slope, it loses most of its moisture but is still warm. It is dynamically heated further as it descends, and so blows down the valleys as a very warm, dry wind. It descends the northern slopes as a dry, hot wind, following relief lines such as the valleys of Upper Rhine, the Aar, the Reuss, and the Rhone from Martigny to Lake Geneva. It melts snow rapidly, often causing widespread avalanches and temperature can rise 10°C in a few hours. It is so dry that danger of fire is serious, and in some Alpine villages, smoking out-of-doors is prohibited during fohn weather.

It may raise the temperature by 10°C to 12°C in a few hours. Consequently, the snow is melted, vegetation, houses, etc., become excessively dry and it results into hazardous avalanches on the leeward slopes. The maximum frequency of the fohn in northern Switzerland is in spring (March to mid-May), when it is useful in melting the winter snow from the pastures. In autumn (October-November), the season of the second greatest frequency, it is useful in ripening the crops, especially the grapes. The fohn blows in the direction of the mountain valleys, but is usually a southerly wind and is strongest in valleys in north-south direction. This wind is characterised by a combination of heat and excessive dryness, together with strong, gusty winds. Such a weather causes physiological as well as psychological reactions: irritableness, headaches, etc., are quite common.

The fohn also dries out the land, trees and twigs, and creates favourable conditions for forest fires. A similar type of wind occurs in all mountainous regions which are affected by depressions. Berg (South Africa), chinook (Rockies), norwester (New Zealand,

Australia), samun or samoon (Iran), santa ana (California, USA) and zonda (Argentina) resemble to fohn wind.

Chinook: 'Chinook' means *'snow-eater'*. It is a warm, dry, southwesterly adiabatic wind which blows down the slopes of the Rockies in parts of USA (Colorado, Wyoming, Montana, North Dakota, Oregon, Washington), and Canada (Alberta, Manitoba, Mackenzie). The word 'chinook' has been derived from the name of a Red Indian tribe that formerly lived near the mouth of the Columbia River (USA).

This wind normally blows on the southern side of a depression which is moving eastward across the continent, and is southwesterly, though its direction is considerably modified by local topography; it thus generally blows during winter and spring (December-April). Being dried and warmed adiabatically by descent from the Rockies, the chinook raises the atmospheric temperature, sometimes by about 20°C (30°-40°F) in only 15 minutes, providing a great contrast to the anticyclonic cold. It melts and dries up the winter snow, and so makes grazing possible almost throughout the winter. It is thus of paramount economic importance, especially in the pastoral regions, from southern Colorado (USA) as far north as the lower Mackenzie River (Canada). Strong and frequent chinooks signify that winter is mild, and pastures are available for grazing practically without interruption. Absence of chinook means a severe winter, and probably heavy losses of livestock.

Chinooks and foehns are intensified considerably if there is ascending motion with precipitation on the windward side of the mountain. While the air ascends from A to B on the windward side, it cools according to the wet adiabatic rate (about 6°C/km). However, when it descends from B to A on the leeward side, it warms up dry-adiabatically (10°C/km). In this manner, the leeward side benefits from the liberation of the latent heat on the windward side. The main reason for the warmness in the lee is, however, that the descent in the lee is far greater than the ascent to the windward of the range.

Norwester: Norwester is a strong wind blowing from the New Zealand Alps, across South Island with fohn-like characteristics.

In the Ganga-Brahmaputra plains of India, a type of squall, usually accompanied by violent thunderstorms and heavy rains and hail-showers, is experienced during the hot season (April to June). Since it comes in Bengal, Assam and Bangladesh at the time of ripening of mangoes, it is also known as *mango-shower* and *Kal-Baisakhi*. At this time, the rainfall of West Bengal, Assam, Bangladesh and Myanmar (Burma) is largely derived from the norwester. In Assam and northern parts of West Bengal these winds are extremely useful for tea and jute crops.

Samoon (Samun): An Iranian term for a hot, dry wind of fohn-like character. Derived from the Persian language, the samoon wind descends from the highlands of Kurdistan. At its occurrence, the temperature may rise by 10°C which helps in the melting of ice and snow in the months of February, March and April. It also blows in the autumn season but with low frequency and moderate temperature increases. It damages the vegetation in spring.

Santa Anna: This local wind blows in California (USA), which has the direction from northerly to easterly. Blowing from the desert and having been heated by descent from the mountains, it is hot, dry and dusty, and resembles to fohn and chinook. Like the fohn and sirocco of the Mediterranean Sea, it causes much discomfort to the inhabitants of California. Moreover, it causes considerable damage to crops by drying up the fields and vegetation. It is mainly experienced in winter, being due to passage of depressions but it causes the worst destruction if it occurs in spring when the fruit trees are in blossom, young buds are formed and the flowers are in the fruiting stage. It is often laden with fine sand, which penetrates even small openings and makes life miserable indoors as well as in open. On occasions it has played havoc with anchored ships.

Aandhi (Dust Storm): It normally occurs in the western parts of India in the hot weather in the late afternoon. At the occurrence of a dust storm the visibility is reduced to about five metres. It is always accompanied by violent winds with very little or no rainfall as the moisture is evaporated before reaching the earth surface.

The dust storms, associated with cumulus, cumulus-nimbus clouds, attended by powerful squalls of short duration, range from a few minutes to about half an hour or so, in which individual gust may attain a velocity of over 200 kmph. These dust storms damage the mango crop and uproot the trees causing heavy damage to crops, animals and property.

Dust Devil: It mainly blow in the desert of Sahara. They may also be observed in other deserts of the world. It is a local whirl of dust, usually not more than a few metres in diameter, in which the particles are swept round and round the centre, and are lifted to considerable heights — sometimes to 600 to 900 metres (2,000-3,000 feet) above the earth surface. It is caused by excessive local heating by the sun in an arid region, a strong convection current being formed. It usually moves over a desert at the speed of 10-25 kmph, though speed of over 50 km (30 miles) per hour is sometimes reached, and as many as five or six of them may often be counted at a time by the observer. Being so limited in size, it is harmless even to aviation, unlike the dust storm.

Haboob: Haboob is a type of dust storm experienced in north and northeast Sudan, being most frequent near Khartoum. It generally blows in the afternoon or evening. It occurs mainly between May and September, but is possible at any time of the year. It is usually accompanied by a sudden increase in wind strength and a change of direction, a marked fall in temperature, and an extremely poor visibility caused by the raising of the dust. It is often followed by heavy rain and sometimes thunderstorms. It has great resemblance to the *aandhi* (dust storm) of Rajasthan, Gujarat, Haryana, Punjab and western Uttar Pradesh.

Karaburn: The hot northeasterly wind experienced in Tarim Basin of Xinjiang (China); when the interior of Asiatic landmass is strongly heated, it sets in during the early spring in Central Asia, and continues by day till the end of summer. It is strong, often blowing with gale force, and sweeps up clouds of dust from the desert, darkening the air and causing acute discomfort. The coarse sand is not carried beyond the deserts, but the lighter dust particles are transported to long distances, cause a characteristic haze, and on settling on the ground form loess. The sand blown along by the Karaburn is one of the principal causes of the changes in the course of rivers through the desert.

Shamal: The extremely constant northwesterly wind, experienced chiefly during the summer in Iraq, blows across the Tigris-Euphrates plains. Being uninterrupted by depressions, it blows with great regularity. It is particularly strong during the day, when it carries clouds of dust and causes severe dust storms, especially in southern Iraq, but there is normally a marked lull at night. It is also experienced in winter, but with far less regularity.

Siestan: It is a strong northerly wind experienced in Siestan – a region of eastern Iran. Its speed sometimes exceeds 110 km. It is also known as the *wind of 120 days.*

Simoom: 'Simoom' is an Arabic term for an excessively hot, dust-laden wind (49°-59°C/120-135°F) in the north Sahara, accompanying an occasional cyclonic storm which invades the desert. It appears largely as a swirling phenomenon with some affinities to a whirlwind or *'dust devil'*, but is of longer duration. It is most characteristic of the summer months.

Description of Monsoon

The term *'monsoon'* has been derived from the Arabic word *mausim* meaning 'season'. The Arabs apply it to the seasonal winds of the Arabian Sea, which blow for about six months from the southwest and six months from the northeast. It is now applied generally to the type of wind system in which there is complete or almost complete *reversal* of prevailing direction of winds from season to season. It is especially prominent within the tropics on the eastern sides of the great landmasses, but it occurs, too, outside the tropics, for in Eastern Asia it reaches as far north as about 60°N. South East Asia is pre-eminently a typical monsoon region. Here, the intense heating of the land in summer causes a low pressure area to be established over northwest India, and the in-blowing winds of the southwest monsoon, having originated as the southeast trade winds in the southern Indian Ocean, are warm and saturated with moisture. These winds advance northward over the region from spring to mid-summer, bringing copious rains to large areas, especially where mountain ranges enhance the effect, e.g., on the slopes of mountains of western India and Myanmar (Burma), the eastern Himalayas, the hills of northeast India, the Gangetic plains, the Western and Eastern Ghats, etc.

The Methods

The southwest monsoon moves forward with a definite front, and arrives at each place at approximately the same date each year. The first sudden rain of monsoon is known as the *burst of the monsoon.* Violent tropical cyclones occur in its van and rear periods and occasionally while it is at its peak.

The summer monsoon lasts from the ending part of April to September, though its duration anywhere depends upon the geographical situation of the place in question.

In autumn (October-November in the Northern Hemisphere), as the land cools, higher pressure is established over the continent of Asia, with lower pressure to the south. The northeast (winter) monsoon, which now advances southwards as the southwest monsoon retreats, is thus to most of the region a cold, dry wind, bringing rain only to the limited areas which it reaches by a partial sea track, e.g., the Eastern Ghats (southeast India) and Sri Lanka. The prevailing direction of this monsoon in North China is northwest, and in Central China north, and owing to the proximity of deserts, the winds are dusty as well as dry and cold. The northern India (Indo-Gangetic plain) is protected by the Himalayas and the Tibetan plateau. In the Gangetic plain, the winds are much lighter than those in China.

The topography and relief of land largely controls the effect of the winter as well as the summer monsoons. The winter monsoon lasts approximately from October to March. Other areas which experience similar but less pronounced seasonal changes of wind direction include the southeastern USA, east sides of Central and South America, the Caribbean Islands, Malagasy (Madagascar), East Africa, the Guinea coast of West Africa, Philippines, and North Australia. These are sometimes rather misleadingly termed 'monsoon regions'. In southeast USA, there is not a complete reversal of wind; in the southern continents there are damp, in-blowing winds on the east coasts during the summer but the extratropical landmasses are too small to develop anticyclones of the Asiatic type, and there is no marked system of out-blowing winds.

Natural Soils

The loose material or the upper layer of the mantle rock *(regolith)*, consisting mainly of very small particles, is known as *soil*. The naturally occurring soil is influenced by *parent material, climate, relief,* and the *physical, chemical* and *biological agents* (microorganisms) in it.

Soil mainly consists of mineral particles, a certain proportion of decayed organic material, soil water, soil atmosphere, and living organisms, which exist in a complicated and dynamic relationship with one another.

Basic Materials

The parent material includes both hard, resistant rocks such as granite and slate, and also less resistant rocks such as recent volcanic lavas and ashes, and most of the sedimentary rocks (sandstone, clay, silt and limestone). The term 'rock' is strictly applied not only to granite, sandstone and the like, but also to gravel, clay and unconsolidated sand. The loess and alluvium are the less compact and less resistant soils.

Limestone, for example, is a parent material (rock), Development of soil on a limestone rock is a very slow process. Moreover, most of the soils developed on limestone are thin and dry, and on steep slopes soil may even be non-existent. Several major soil types are recognised which develop on limestone, including red and brown on terra rossa and rendzina on chalk.

Constituents of Soil

A soil is made up of four elements: inorganic or mineral fraction (derived from the parent material), organic material, air and water. The abundance of each component and its importance in the functioning of the soil system vary from horizon to horizon and from one soil to another.

Humus: The end-product of the breakdown of dead organic material is known as *humus* – a structureless, dark-brown or black jelly found beneath the soil surface. In uncultivated land, the humus is derived from the natural decay of previous generations of plants, while in the ploughed and cultivated land it is supplied as some kind of manure. The humus of ordinary soil is black, and is thus responsible for making the soil darker than the subsoil. It plays an important but very complicated part in maintaining the fertility of soil. The amount of humus in different soils varies considerably; some, like the peat soil, consist largely of slightly decomposed organic matter which has not yet become humus.

Soil Texture: A soil is generally characterised by the size of its particles. A clayey soil may thus be described as fine, a sandy soil as coarse, while a silty soil is intermediate. If one handles a moist soil sample of each of these he feels gritty, sticky and silky, respectively. The standard unit for the measurement of soil particles is the millimetre, but a smaller unit is the micron (1 micron= 0.001 mm), which is applicable, for instance, to the measurement of soil colloids.

Sandy Soil: A light soil consists mainly of sand, i.e., grains of quartz with considerable air spaces between them. The sand may either be 'coarse' where the particles are between 0.2 and 2 mm in diameter, or 'fine' where the grains between 0.05 and 0.2 mm are just visible to the naked eye. These light soils allow water to drain through rapidly, taking soluble plant foods with it. They are therefore 'hungry' soils, which not only need constant manuring but may dry out completely during a period of drought so that shallow-rooted crops fail and pastures 'burn'. They are good for horticulture (vegetables and fruits), legumes *(moth)*, groundnut and *bajra* (bulrush-millet).

Clayey Soil: This is an exceptionally fine grained soil, very retentive of moisture. It often becomes plastic when mixed with water. The individual grains of clayey soil are 0.002 mm in diameter. These particles consist mainly of hydrated aluminium silicates. A clay contains little air and can hold more water, so forming a sticky mass, but when it dries out completely, it forms a hard, concrete-like surface, seamed with numerous cracks. Sometimes, a compacted solid layer of clay in the subsoil is formed, which is known as *claypan,* and is often hard and difficult to dig or plough.

Clayey soils are often rich in plant food and give much better yields than that of sandy soils. They are devoted to rice, perennial grasses and clover. Efficient drainage methods, modern machinery and careful liming enable clayey soils to grow roots, green crops and cereals.

Silty Soils: Silty soil is finer than sand but coarser than clay. Its particles are assumed to have a diameter between 0.02 and 0.002 mm. These soils are rich in humus contents and are devoted to numerous cereal and non-cereal crops.

Loamy Soils: It is highly fertile soil consisting mainly of a mixture of sand and clay, together with silt and humus. It has the good qualities of both sand and clay, but not their bad qualities. It comprises an almost equal mix of sand and silt with less than 30 per cent clay. It can retain some moisture and plant food even under the adverse weather and climatic conditions. It is well-aerated and drained, and can be readily worked. It is generally devoted to wheat, barley, legumes, sugarcane, sugarbeet, maize, millets, rice, grasses, vegetables and orchards.

Soil Acidity: In cool and moist areas, percolating groundwater leaches out the soluble bases (calcium). As a result, the soils gradually become lime-deficient which increases the acid and 'sour'.

The acidity and alkalinity of soils is expressed in the pH value, which is a scale measured in terms of the hydrogen ion concentrations held by the soil colloids (particles). In pure water, one part in 10 millions is dissociated into hydrogen ions, i.e., 10^{-7}, and pH is 7; this is a neutral state on the scale of acidity. If a strong alkali such as caustic soda is dissolved in water, the solution is marked as alkaline (pH 14). By contrast, hydrochloric acid has a pH value of 3. A neutral soil has a pH value of about 7.2, an acid soil of less than 7.2 (sometimes as low as 3), and a strongly alkaline soil of about 8 or higher.

Both the highly acidic and alkaline soils are injurious to crops. If the soil becomes unduly acidic, the farmers apply lime in various forms to meet the requirements of the soil. In practise, a pH value between 6 and 6.5, i.e., very slightly acidic, is desired. Lime not only helps to neutralise the excess acids and so 'sweeten' the soil, but it also encourages bacteria and helps to improve the physical texture of heavy soils.

High soil acidity is typical of cold, humid climates. In arid climates, soils are typically alkaline. Acidity can be corrected by the application of lime, a compound of calcium, carbon and oxygen ($CaCO_3$), which removes acid ions and replaces them with the base calcium.

Soil Air: The air content of a soil is vital, both to itself and to organic life within it. A certain amount of air is contained between the individual particles except for the waterlogged soils. The air in the soil helps in the process of oxidation which converts part of the organic material into nitrogen in a form readily available to the plants. On the other hand, too high degree of oxidation (in the tropical lands) may consume so much organic material that the soil becomes increasingly sterile.

Moreover, most bacteria, present in the soil in infinite numbers, require oxygen and are said to be *aerobic.* As these organisms are partly responsible for breaking down plant remains, absence of air limits their activity. Earthworms too have an important process.

Soil Water: Depending on the texture of the soil, water moves downward by percolation. The amount of water in the soil varies from almost nil in arid climates which makes life virtually impossible for organisms, to a state of complete waterlogging which excludes all air, causes a reduction of bacteriological activity, and limits decomposition. In damp climates, especially in high latitudes where the evaporation rate is low, water tends to move predominantly downward, particularly in coarse-grained sandy soils.

Soil Acidity and Alkalinity

pH	4.0 4.5 5.0	5.5	6.0	6.5 6.7	7.0	8.0	9.0	10.0	11.0
Acidity	Very strongly acid	Strongly acid	Moder-ately acid	Slightly acid	Neutral	Weakly alkaline	Alkaline	Strongly alkaline	Excessively alkaline
Lime require-ments	Lime needed except for crops requiring acid soil	Lime needed for all but acid tolerant crops		Lime generally not required	No lime needed				
Occurrence	Rare Frequent	Very common in cultivated soils of humid climates			Common in sub-humid and arid climates			Limited areas in deserts	

This dissolves the soluble minerals in the soil, together with soluble humus material and carries both downward, a process called *leaching* or *eluviation.* A typical leached soil is known as *podzol,* a Russian word meaning 'ash' because the surface layer is often greyish or ash-coloured. In a hot, arid climate, evaporation exceeds precipitation for greater part of the year, so the water tends to move upward and the soil dries out. Consequently, in some areas, a thin salty layer is formed on the surface. This process of salinisation can produce an extremely saline soil known as *soloncbak (reh* or *kallar).*

Soil Horizons: A layer of soil which lies more or less parallel to the surface and has fairly distinctive soil properties is known as *soil horizon.* Most of the soils possess soil horizons, having distinctive layers that differ in physical composition, chemical composition, organic content or structure. Soil horizons are developed by the interactions through time of climate, living organisms, and the configuration of land surface. These are often distinguished by their colours.

Soil horizons are of two types: organic and mineral. *Organic horizons,* designated by the capital letter 'O', overlie the mineral horizons and are formed from accumulations of organic matter derived from plants and animals. Soil scientists recognise two possible layers: *Oi horizon* containing decomposing organic matter that is recognisable as leaves or twigs, and *O_a horizon* containing material that is broken down beyond recognition by eye. The material is humus.

Mineral horizons lie below the organic horizons. Four main horizons are important — A, E, B and C. Plant roots readily penetrate A, E and B horizons and influence soil development within them. Soil scientists limit the term 'soil' to refer to the A, E and B horizons.

The *A horizon* is the uppermost mineral horizon. It is rich in organic matter, consisting of numerous plant roots and downward humus from the organic horizons above. Next is the *E horizon.* Clay particles and oxides of aluminium and iron are removed from the E horizon by downward-seeping water, leaving behind pure grains of sand or coarse silt.

The *B horizon* receives the clay particles, aluminium and iron oxides, as well as organic matter washed down from the A and E horizons. It is made dense and tough by the filling of natural spaces with clays and oxides.

Beneath the B horizon is the *C horizon,* which is not considered part of the soil. It consists of the parent mineral matter of the soil. Below the regolith lies the bedrock or sediments of much older age than the soil.

Soil's Profile

A vertical section of soil in which all the soil horizons are shown is known as *soil profile.* Most of the soil profiles include three master horizons.

Soil's Fertility

Soil fertility is the ability of soil to sustain plants. Soil has fertility when it contains organic substances and clay minerals that absorb water and certain elements needed by plants. Most of the soils have, however, been threatened by soil erosion. Increased crop production from artificial fertilizers and new crop hybrids partially mask their effect, but such compensations are nearing the end of their capacity. Soil depletion and loss are at record levels all over the world. The impact on society is potentially disastrous as population and food demands increase.

Soil science is an interdisciplinary subject, involving physics, chemistry, biology, mineralogy, hydrology, taxonomy, geography, climatology, and cartography. Physical geographers are interested in the spatial patterns formed by soil types and the physical factors that interact to produce them. As an integrative science physical geography is well suited for this task.

Soil science must deal with a substance whose characteristics vary from kilometre to kilometre. Thus people interested in soil need information specific to their area. In many locales, an agricultural extension service can provide this information and perform a detailed soil analysis.

Forms of Soil

Soils are divided into: i) zonal, ii) intrazonal, and iii) azonal categories.

Zonal Soils: A soil whose characteristics are dominated by the influence of climate and vegetation is known as a *zonal soil.* These soils occur on gently undulating land where drainage is free and where the parent material is of neither extreme texture nor chemical composition. They occur in latitudinal zones. Zonal soils are further divided into: i) pedalfers, and ii) pedocals.

Intrazonal Soils: A soil which has been influenced in its development less by climate and vegetation than by other local factors, such as defective drainage, excessive evaporation or an unusual parent material (such as limestone), terrain or age. They am be sub-divided into: i) calcimorphic soils, ii) halomorphic soils, and iii) hydromorphic soils (bog-peat, fen-peat or meadow soil).

Azonal Soils: A soil which has not been sufficiently subjected to soil-forming processes for the development of a mature profile and so is little changed from the parent rock material. Azonal soils do not have B horizon because it is too immature. Thus, the A horizon lies immediately above the C horizon. Examples are soil forming on screes, recently deposited alluvium, sand dunes, newly deposited glacial draft, wind-blown sand, marine mud flats and volcanic soils. Azonal soils are further divided into: i) lithosol, ii) regosol, and iii) entisol (alluvial).

Zonal Soils

Tundra Soils: Tundra soils are closely influenced by the low temperatures, a slow rate of evaporation throughout the year, and a brief growing season for plants. Chemical weathering and biological activity in such a cold climate are very limited and waterlogging is widespread in summer. In areas of better drainage a more mature soil, known as *Arctic brown soil*, may develop on sandstone.

Podzol: Podzol is a Russian term, meaning 'ash soil'. It occurs widely in the moist cool climates of mid-latitudes. It develops under a natural vegetation of coniferous forests. It is found in the areas, where the precipitation is only moderate but evaporation is limited. It is a poor agricultural soil. Huge areas of the coniferous forest regions of northern Canada and the northern Russia are covered with podzols.

Grey-Brown Podzols: It is a soil intermediate between brown earth and the true podzols. It is characterised by dark surface. It develops in a temperate climate of moderate rainfall but it has a greater organic content and is less leached than a true podzol.

Brown Earth or Brown Forest Soil: These soils, typical of much of the area formerly covered by the deciduous forests of Europe, are rich in organic matter. Agriculturally, these are much superior to podzols. Most of these soils have been cultivated for centuries. The natural fertility of these soils has depleted, and therefore manuring is essential.

Chernozem: In Russian language black earth is known as *chernozem*. It is a fine, fertile soil, black or dark brown in colour, covering an extensive area of Russia, Ukraine and Kazakhistan, and parts of Hungary, and Romania. The soil is so rich in plant foods that it will take crops for long period without the addition of fertilizers. It is well known for the cultivation of cereal crops. It is also found in Canada and USA, sprawling from Saskatchwan (Canada) in the north to Texas (USA) in the south. It is associated with natural cover of grass.

Brunizem (Prairie Soils): It is a dark-coloured soil occurring in the mid-latitude sub-humid climatic zone beneath the temperate grasslands. Prairie soils form in continental interiors, where rainfall varies between 60 and 100 cm annually. These are transitional soils between pedalfers and pedocals. The tall grasses stimulate a rich humus content. These soils occur in eastern Europe, and in the USA from southern Minnesota to Oklahoma.

Chestnut-Coloured Soil: These soils are essentially a variety of chernozem, modified by greater aridity. These occur in the steppe areas where about 20 to 25 cm of rainfall is recorded annually. These are loose and friable, containing much humus because the natural vegetation is in the form of grasses. These soils are found in the drier parts of Russia, Romania and Hungary. These are also found in the high plains of the USA and the drier parts of Argentina pampas and the South African velds.

Cool Desert Soils: It is a group of soils sometimes known as *sernozem*. It occurs in the more arid areas of the middle latitudes. The humus content is small. The upper horizon is light grey. It is found in Turkmenistan (east of the Caspian Sea), and parts of western USA. These soils are rich in plant foods, and wherever irrigation water is available, they give good returns.

Latosol: It is a lateritic soil which develops in the hot and humid areas, with potent evaporation. In the humid tropics, the chemical weathering and leaching are intense, leading to deep soil profiles of free drainage. These soils, yellow to red in colour, are rich in hydrated oxides of iron, aluminium and manganese. Their texture varies from a kaolinitic clay to a loamy sand. It has poor vegetation cover and is deficient in humus. These soils are found in Amazon and Zaire basins, Central America, the east coast of Brazil, eastern Madagascar and Indonesia. These are also found in Sri Lanka, Myanmar (Burma), Vietnam, and East Indies. They need heavy manuring and fertilisation as they are rapidly exhausted.

Tropical Black Earth: It is a dark-coloured, heavy textured soil, characterised by swelling during the rainy season but drying out and cracking during the dry periods. In Peninsular India, it is known as *regur soil* (black cotton soil).

Red Desert Soils: These are reddish-brown soils of the hot deserts. These are mainly of a sandy texture and contain much salt because of the absence of leaching.

Intrazonal Soils

Saline Soils: There are certain soils which contain considerable quantity of soluble salts. Such soils are known as *saline soils.* The pH value of saline alkali soils is around 8.5. These are of widespread occurrence wherever there is a sufficient degree of evaporation. The strong salt solutions rise by capillarity and form a greyish surface crust, below which is a granular salt-impregnated horizon. A typical saline soil is called *solonchak.* If there is a rather higher rainfall some surface salts may be leached out, so that the saline layer occurs lower in the B horizon. Such soils are also known as *solontez.*

Peat Soils: It is an unconsolidated black or dark-brown soil material consisting of slightly decomposed or undecomposed fibrous vegetable matter that has accumulated in a waterlogged environment in which air is virtually absent. It is formed under cool, humid climatic conditions. Peat is a very important source of fuel in some northern countries, e.g., Russia and Ireland, but it gives less heat. Diagenesis of peat leads to coal and lignite formation.

Calcareous (Calcimorphic) Soils: Soils containing sufficient calcium carbonate are known as *calcareous soils.* The parent material for these rocks is limestone. *Rendzina,* red and brown Mediterranean soils, and the brown calcareous soils are some of the examples of this group.

Azonal Soils

Mountain Soils: These soils develop on stony unstable surfaces as scree slopes or on glacial moraines. The soil may consist to a large extent of rock fragments, which disintegrate physically mainly by frost action, but suffer little chemical change. These are also known as *scree soils.*

Alluvial Soils: These soils are derived from a mixture of sand, silt and clay, consisting of well-mixed rock waste transported and deposited in level beds by running water, and replenished at the times of flood. Some of the most fertile lands in the world consist of alluvium deposited in the valleys and deltas of the great rivers. The Indo-Gangetic plain, the plains of Hwang-Ho and Yangtze-Kiang, the plains of Mississippi and Missouri, etc., are some of the examples of alluvial soils. The alluvial soils are intensively cultivated.

Marine Soils: These soils are deposited by sea-wave action. Materials derived from marine origin are built up along low-lying coasts in the form of mud banks, sand banks and dunes by natural processes, stimulated by artificial reclamation. Along the coast of Belgium, Netherlands, Germany and Denmark areas of former sea floor, known as *polders,* are enclosed by dykes and then drained which produce various soils-like marine clays.

Glacial Soils: These soils are deposited by the glacial and fluvial-glacial processes. The variety of glacial deposits includes tills, outwash sands, and gravels. These soils are generally poor in humus content.

Wind-Blown Soils: The wind deposits sand sheets, dunes and loess which are the wind-blown soils. These soils are generally poor in humus content. The *loess* deposits of China, *horde* of Germany, *limon* of Belgium and northeastern France are all fine textured.

Volcanic Soils: The recent volcanic activity deposits lava, ash and pumice, which weathered easily, are transported lower down the slopes of volcanoes by rain-wash and torrents. The new higher lava forms bare grey sheets, but the lower weathered materials are immensely fertile. Villages and vineyards cling around the slopes of Vesuvius and Etna, in spite of repeated disasters in the past and the ever present threat of volcanic activity.

Soil's Classification

Soil classification is a cumbersome task. A number of different classification systems are in use worldwide. The United States, Canada, the United Kingdom, Germany, Russia, Australia, China, India, and the United Nations Food and Agricultural Organisation (FAO) each has its own soil classification system.

The Russians contributed greatly to modern soil science because they were the first to consider soil as an independent natural body with a definite individual soil genesis. By contrast, the American and European scientists still consider soil as a geologic product, or simply a mixture of chemical compounds. The Russians determined that broad interrelationships exist among the physical environment, vegetation and soils.

The first formal U.S. soil classification system was attempted by C.F. Marbut (1930). Subsequently, the US soil conservation service developed a new classification of soils in 1975. This is known as the *US soil taxonomy.* In this classification the soils of the world have been divided into eleven orders.

Oxisol (Tropical Soils): Oxisol soils are largely confined to hot and humid climates of the equatorial region. The intense temperature, persistent moisture and almost uniform daylength of equatorial latitudes greatly affect soils. These soils have a distinctive horizon with a mixture of iron and aluminium oxides. The equatorial and tropical rainforests which grow in them luxuriously add humus content in the soils.

The oxisols are generally reddish and yellowish as the humus content is leached under the impact of heavy precipitation. The sub-surface horizon in an oxisol soil is highly weathered, containing iron and aluminium oxides. If these horizons are subjected to a repeated wetting and drying, a hardpan (hardened soil layer) develops, called *plinthite* (Greek 'brick'). This form of soil is called a *laterite.* The laterite soil can be quarried in blocks and used as a building material. In the areas of isolation and relative isolation, these soils are still utilised for shifting cultivation. Some of the large plantations (rubber in Malaysia, Indonesia) are in the oxisols.

When oxisols are disturbed, soil loss can exceed a thousand tons per square kilometre a year. The regions dominated by the oxisols and rainforests are the focus of much worldwide environmental attention.

Aridisols (Desert Soils): Occupying about 19.2 per cent of the earth's land surface, aridisols are confined to the dry regions of the world. These soils have pale, light soil colour near the surface.

The aridisol regions have marked periods of soil moisture deficit and generally inadequate soil moisture for plant growth. The high potential evapotranspiration and low precipitation produce very shallow soil horizons. Aridisols also lack organic matter of any consequence.

These soils are characterised by salinisation. Salinisation results from excessive potential evapotranspiration rate in the deserts and semi-arid regions of the world. Salts dissolved in soil water are brought to surface horizons and deposited as surface water evaporates. These deposits kill plants when the salts accumulate near the root zone. In the Nile, Indus and Ganga River valleys, aridisols are intensively cultivated. Unfortunately, substantial tracts of productive aridisols have been rendered useless from agricultural point of view simply due to careless management of irrigation water.

Mollisols (Grassland Soils): Mollisols, which contain large humus content, are some of the most productive soils of the world. The surface layer of these soils is dark in colour. They are soft, even when dry, with granular or crumbly texture. These humus-rich organic soils are high in basic metals (calcium, magnesium, and potassium). In terms of water balance, these soils are intermediate between humid and arid soil moisture.

Soils of the steppes, prairies, pampas, velds and downs belong to this soil group. These soils are also found in Manchuria (China), Ukraine, Russia, and Europe. Agriculture ranges from large scale commercial grain farming to grazing along the drier portions of the soil.

In some of the mollisols, calcification process is found. Calcification is an alleviated accumulation of the calcium carbonate or magnesium carbonate in the B and C horizons. When cemented or hardened, these deposits are called *caliche,* or *kunkur.* These occur in widespread soil formations in central and western Australia, the Kalahari region of Africa, the high plains of west-central USA, and the Indo-Gangetic plain.

Alfisols (Moderately Weathered Forest Soils): Alfisols are the most widespread of the soil order, extending from near the equator to high latitudes. Most alfisols have a pale, greyish brown to reddish colour and are considered moist versions of the mollisol soil group.

Alfisols are fertile, which depends on the availability of moisture and temperature. These soils are usually supplemented by moderate application of lime and fertilizers in areas of intensive agriculture. In the subtropics and temperate latitudes they are also devoted to grape, almond, walnut, fig, olive and citrus fruits.

Ultisols (Highly Weathered Forest Soils): Ultisols are found in the subtropical forests which are highly weathered. The soils tend to be reddish because of residual iron and aluminium oxides in A horizon.

The increased precipitation in the Ultisol regions means greater mineral alteration, more alluvial leaching, and therefore, a lower level of basic metals (calcium, potassium, and magnesium), leading to infertility. Fertility is further reduced by certain agricultural practises and the effect of soil damaging crops such as cotton and tobacco, which deplete nitrogen and expose soil to erosion. However, these soils respond well if subjected to good management — for example, crop rotation that restores nitrogen and cultivation practises that prevent sheet-wash and soil erosion Much needs to be done to achieve sustainable management of these soils.

Spodosols (Northern Coniferous Forest Soils): Spodosols in the cold and forested wet areas are characterised by coniferous vegetation. These soils are mainly found in North America (northern USA, northeast Canada), Eurasia, Denmark, the Netherlands, eastern Sweden, Finland and southern England. Because there are no comparable climates in Southern Hemisphere, this soil type is not identified there. Spodosols form from sandy parent materials, shaded under evergreen forests of spruce, fir and pine. Spodosols with more moderate properties form under mixed or deciduous forests.

Spodosols lack humus and clay in A horizon. An eluviated white horizon, sandy and leached of clays and iron, lies in the A horizon instead and overlies a horizon of eluviated organic matter and iron and aluminium oxides. The surface horizon receives organic litter from base-poor, acid-rich trees, which contribute to acid accumulation in the soil. These are ash-grey in colour. Good agriculture is possible only if nitrogen, phosphate and potash (potassium carbonate) are added to the soil and a scientific rotation of crops is adopted by the farmers.

Entisols (Recent, Undeveloped Soils): Entisols are the undeveloped soils in which the vertical development of horizons is missing. They occur in many climates. Entisols are true soils that have not had sufficient time to generate the usual horizon.

From the agricultural point of view entisols are generally poor, although those formed from river silt deposits are quite fertile. These soils are found in active slopes, alluvium-filled floodplains, poorly-drained tundra, tidal mad flats, sand dunes, erg (sandy deserts), and plains of glacial outwash.

Inceptisols (Weakly Developed Soils): Inceptisols are inherently infertile. These are weakly developed young soils, although more developed than the entisols. Inceptisols include a wide variety of different soils, all holding in common a lack of maturity with evidence of weathering just beginning. Inceptisols are associated with moist soil regimes and have no distinct alluvial horizon.

Inceptisols include the soils of most of the Arctic tundra, glacially derived till and outwash materials of Siberia, New York, and alluvium of the Mekong, Irrawady and Gangetic floodplains.

Andisols (Volcanic Parent Material): Andisols occur in areas of volcanic activity. These are derived from volcanic ash and glass. The Andisols of Hawaii are well known for the high yields of sugarcane and pineapple. Their distribution is small in areal extent. They are mainly found in the *Ring of Fire* in the Pacific Rim.

Vertisols (Expandable Clay Soils): Vertisols are heavy clay soils. They contain more than 30 per cent swelling clays (which swell significantly when they absorb water). They are located in regions experiencing highly variable soil moisture balances through the seasons. These soils occur in areas of sub-humid to semi-arid moisture and moderate to high temperature. Vertisols frequently form under savanna and grassland vegetation in tropical and subtropical climates and are sometimes associated with a distinct dry season following a wet season.

Vertisol clays are black when wet and range to brown and dark grey. These deep clays swell when moistened and shrink when dried. In the process, vertical cracks develop, which widen and deepen as the soil dries, of the size 2-3 cm wide and upto 40 cm deep. Vertisols are high in bases and nutrients and thus are some of the better farming soils wherever they occur. They are found in the Deccan plateau, Madagascar, Mozambique, Namibia, eastern coast of Brazil, Uruguay, Texas and Argentina. These are generally utilised for the cultivation of cotton, millets and maize.

Histosols (Organic Soils): Histosols are formed from accumulation of thick organic matter. In the mid-latitudes, when conditions are favourable, beds of former lakes may turn into histosols, with water gradually replaced by organic material to form a bog and layers of peat. Histosols also develop in small, poorly drained depressions where conditions are ideal for significant deposits of sphagnum peat to form.

This material can be cut, baled and sold as a soil amendment. Dried peat has served for centuries as a low grade fuel. The area southwest of Hudson Bay in Canada is a typical histosol formation area.

Soil Erosion

The removal of soil at a greater rate than its replacement by natural agencies (water, wind, etc.) is known as *soil erosion.* Some soil erosion occurs without the intervention of human activities but the later often accelerate the natural processes. Soil erosion may be divided into four main types:

1. Wind erosion,
2. Sheet erosion,
3. Rill erosion, and
4. Gully erosion.

Forms of Erosion

Wind Erosion: Wind is a very important agent of soil erosion. The wind removes the dry, unconsolidated material. Wind erosion is a serious problem in the arid and semi-arid climates. The main remedy of wind erosion has been to keep as much of land as possible covered with vegetation (cover crops). Another method is the planting of long lines of trees as wind breaks. Wind erosion is quite serious in almost the entire arid and semi-arid regions. The state of Rajasthan, western Haryana, southern Punjab, northwestern Madhya Pradesh and Marathwada (Maharashtra) are seriously affected by wind erosion.

Sheet Erosion (Sheet Wash): It is widespread removal of surface debris. It occurs because of the overland flow on low gradient slopes, generally at relatively slow speeds and over long periods. It comprises the two processes of rain drop impact and overland flow, with sheet wash carrying away the particles dislodged by the falling rain. When continuously heavy rainfall occurs sheet wash becomes sheet flood, and soil erosion becomes extremely severe as sheet erosion is accelerated. This type of soil erosion is widespread in the humid regions.

Rill Erosion: The removal of surface material, usually soil, by the action of running water. The processes create numerous tiny channels (rills) a few centimetres in depth, most of which carry water only during storms. The head of the rill system may not extend all the way up to the watershed divide, thereby leaving a zone of no rills across which the depth of overland flow is insufficient to develop an erosive force equal to the forces of cohesion which hold the soil particles in place. Rills are, however, more localised and develop in the humid climatic regions.

Gully Erosion: Gully erosion takes place when the stormy rains produce a concentrated run-off thereby removing soil and other poorly consolidated sediments. The main cause of gully erosion is the removal of vegetation, particularly of trees with their widespread binding roots. Typical examples of gully erosion are the Chambal ravines (India), the Tennessee Valley (USA), Italy, Greece, Cyprus, Sicily, etc. The maintenance of an effective vegetation cover may go a long way in reducing the development and process of gully erosion. Improved agricultural practises are also necessary to overcome the problems of gully erosion.

Cycle of Erosion

The 'cycle of erosion' or the 'geographical cycle of erosion' is a concept advocated by W.M. Davis (1889) about the evolution and modification of the physical landscape in which the various stages of erosion are believed to be parts of a cyclic process, which flows an orderly sequence. In the postulation of this concept Davis was influenced by Charles Darwin's Theory of Organic Evolution (*The Origin of Species*, 1859).

In the opinion of Davis "land forms evolve like the evolution of organic life". The cycle of erosion has been defined by him as 'geographical cycle' or a period of time during which an uplifted land mass undergoes its transformation by the process of land sculpture, ending into a low featureless plain — a peneplain. It may be added that the variations in altitude, structure, diastrophic history and climatic conditions may result in notable departures from the sequence.

Analysis of Morphology

The model of 'morphological system' was put forward by the German geomorphologist Walther Penck in 1924 who rejected the Davisian concept of 'geographical cycle'. The typical feature of Penck's model is that the characteristics of landforms of a given region are related to the tectonic activity of that region.

In the opinion of Penck, the landscape is the result of the relative intensity of the degradational processes and the phases and rates of uplift. In other words, landforms were determined by the rate of uplift and the rate of erosion.

Contrary to the Davis' theory that "landscape is a function of structure, process and stage (time)", in the morphological system model, Penck postulated that "geomorphic forms are an expression of the phase and rate of uplift in relation to the rate of degradation". He assumed that interaction between the two factors (uplift and degradation) is continuous.

According to Penck's model, the landscape development (in the words of Davis) begins with the upliftment of *primarumpf* (initial landscape with low height and relief) representing an initial featureless broad land surface.

Penck also assumed that in the beginning the uplift is characterised by exceedingly slow upheaval of long duration and thereafter the rate of uplift is accelerated which ultimately stops after passing through the intermediate phases of uniform and decelerating rates of upheaval. In brief, most tectonic movements began and ended slowly. The initial uplift begins with regional updoming and the landform development passes through the following three phases:

Phase of Waxing (Accelerating) Rate of Landform Development: As stated above, under the impact of endogenetic forces, the land surface rises initially slowly but after sometimes the rate of upliftment is accelerated. In this phase, owing to the upliftment and consequent increase in channel gradient and flow velocity, the streams (rivers) continue to degrade their valleys with accelerated rate of downcutting.

The rate of upliftment far exceeds the rate of valley deepening which results in the formation of the deep and narrow V-shaped valleys and gorges. Since the valley deepening does not keep pace with the upliftment of landmass, the absolute height continues to increase. In other words, the altitudes of divide summits as well as valley bottoms continue to increase as the rate of upliftment far exceeds the rate of vertical erosion.

In this phase, both the maximum altitude (absolute height above the sea level) and the maximum relief (relative height of the valley floors) increase.

Phase of Uniform Development of Landform: On the basis of upliftment and degradation this phase may be divided into three sub-phases:

1. This sub-phase is characterised by still accelerated rate of uplift. The absolute height continues to increase because the rate of erosion is still less than the rate of upliftment. The maximum altitude (absolute relief) is attained but relative relief remains constant because the rate of valley deepening equals the rate of lowering of divide summits. The valley sides are steep.

   This phase is also called the *phase of uniform development*, because of uniform rate of valley deepening and lowering of divide summits.
2. In this sub-phase, altitude (absolute relief) neither increases nor decreases. This is due to the matching upliftment with rate of erosion. The slopes of the valley sides are still straight. This phase is, thus, characterised by constant absolute and relative reliefs. In other words, it is the phase of uniform development of landforms.
3. In this sub-phase, upliftment of land stops completely. Absolute reliefs (altitudes of summit divides) start decreasing. Relative reliefs (valley floors) remain constant because the rate of the lowering of divide summits equals the rate of valley deepening. These processes lead to uniform development of landscape.

Phase of Waning Development of Landscape: This phase is increasingly dominated by the erosional process. The lateral erosion is quite significant, the valley deepening markedly decreases and the valley widens. This phase is marked by progressive decline of landforms. Both the absolute and relative reliefs decline. The valley side slope consists of two segments. The uppermost segment maintains its steep angle which is called as *gravity slope*. The lower segment of the valley is called *wash slope* which is composed of talus materials of lower inclination, formed at the base of valley sides. The advanced stage of the phase is characterised by inselbergs and a series of concave wash slopes. Such extensive surface produced at the end called as *endrumpf* may by considered equivalent to peneplain.

Model of Penck

Penck was a German and his theory about the evolution of landforms could not become popular outside Germany owing to the difficult technical terminology he used in his model. His terminology was *ill-defined, vague, obscure* and at some places *contradictory*. It was because of these reasons that his model was not properly interpreted by the English translators.

Penck's model of evolution of landforms was severely criticised by the American geomorphologists. Penck's concepts of parallel retreat of slope and continued crustal movements were not acceptable to the American experts. His concepts of long continued upliftment and tectonic speculations could not find any support. Penck has drawn our attention to some of the weaknesses of Davis' model, but there are practical difficulties in accepting Penck's concept as an alternative to the concept of Davis. Credit, however, should be given to Penck who has put forward a new viewpoint and has indicated a new possible direction of investigation. His concepts of slope development and weathering processes are thus of much geomorphological significance.

Cycle of Flow

Normal cycle of erosion is synonym for the fluvial cycle, the geographical cycle, the Davisian cycle and the cycle of erosion by river. Fluvial processes are most worldwide spread (covering most parts of the world) and most significant geomorphic agent. Water plays an important role even in the arid and glacial regions. This concept was also advocated by Davis in 1889 who considered humid temperate areas as the most normal case of fluvial cycle.

In the opinion of Davis, the normal cycle of erosion begins with the upliftment of any landmass with reference to sea level. As the land rises, the rivulets and streams are originated and their erosional work starts. In the beginning, the rate of uplift exceeds the rate of erosion. This results into increase in the absolute and relative relief. After some time, upliftment of the land stops and erosion becomes more active. The land area tectonically remains stable, i.e., there is crustal stability for long period of time during which there is neither upliftment nor subsidence of land area. There is progressive development of river valleys in sequential order and the whole land area passes progressively through three successive stages of (i) youth, (ii) mature, and (iii) old. The land is ultimately transformed into a low featureless plain of undulating surface. Thus, the end-product of normal cycle of erosion is peneplain which is marked by undulating surface with residual convex and concave slope and low hills known as *monadnocks*, *unakas* and *mosores*. The main characteristics of the successive stages of normal cycle of erosion are briefly presented in the following paras:

Youth (Adolescence) Stage: This is the first stage, in which a few consequent trunk streams develop. In the beginning, very few tributaries of the consequent stream develop. The slopes are dominated by numerous rills and gullies which increase their lengths through headward erosion. The streams with the help of angular boulders and pebbles do pothole drilling which is the most effective tool of downward cutting. The valleys in the youth stage are steep-sided and at places acquire the shape of gorge or canyon. The water-divides are extensive and wide. The lateral erosion is less significant. There are numerous waterfalls, cataracts and rapids in the river course. River capture is another characteristic of the youth stage. The duration of this stage is, however, short.

Mature Stage: In the mature stage the valleys of the consequent rivers and their tributaries have extended and the region has a well-developed drainage system. The lateral erosion is quite significant. The steep-sided valleys, gorges and canyons are replaced by broad and flat valleys. The rivers deposit boulders, pebbles, shingles and gravels at the foothills which acquire the shape of alluvial fans or alluvial cones. Water-divides are narrowed down due to lateral erosion. The river becomes graded. There are numerous meanders and ox-bow lakes in the floodplain. Deposition of sediments on either side of the valley leads to the formation of natural levees. The depositional processes are more pronounced.

Old Stage: The tributaries to the consequent streams are usually fewer. There is almost complete absence of down cutting. The lateral erosion and valley widening continues. The streams become sluggish due to gentle gradient. The rivers adopt highly meandering courses. The main channel of the river is divided into numerous distributaries and the river becomes braided. Rivers make extensive deltas at their mouth. Lakes, swamps and marshes are the other characteristic features of the old stage. The entire landscape acquires the shape of extensive flat-plain which is called as *peneplain.*

Interruptions in Cycle of Erosion Any type of obstacle in the normal process of cycle of erosion is called *interruption of cycle.* The main causes of interruption of cycle may be either climatic or tectonic or both. The interruption in cycle of erosion caused by positive movement of base level (owing to subsidence of landmass or rise in sea level) shortens the cyclic time as it advances forward the stages of cycle of erosion. For example, if the cycle is in the youth stage, it may advance to mature stage, or if it is in the mature stage, it may advance to old stage. Contrary to this, the negative change in the base level of erosion (due to upliftment of landmass or fall in sea level) lengthens the cyclic time as the cycle is pushed backward, e.g., if the cycle is in the mature stage, the cycle goes back to the youth stage and if it is in the old stage, it recedes to mature stage. Thus, the cycles punctuated by interruptions are called as *interrupted cycles* which lead to occurrences of several cycles in a region. Such cycles are called *poly cycles* in the evolution of landforms. If the poly cycles (multicycles) occur in succession, they are called as *successive cycles of erosion* and the landscapes resulting from them are called *poly cyclic* or *multi-cyclic* landscapes. The Appalachian region of USA and the Chhotanagpur plateau of India present the typical examples of poly cyclic landscapes where several cycles have been completed.

Volcanic Interruptions: The occurrence of volcanic eruptions pour enormous quantities of lava which cover large areas and obstruct surface drainage. Consequently, the existing stage of cycle of erosion is interrupted, and a fresh cycle of erosion starts. Such types of volcanic interruptions have occurred during the geological past (Cretaceous period) over the Indian Peninsula when the lava flows covered vast areas of the Deccan Plateau.

Climatic Interruptions: Climatic interruptions occurred due to major changes in the climate of the concerned region. For example, if the fluvial cycle of erosion in a humid region is passing through mature stage, and there occurs sudden climatic change, resulting into either extremely dry conditions or extremely cold conditions, then the cycle of erosion is interrupted and a new cycle of erosion in the form of arid or glacial cycle commences. Conversely, if the climate becomes more humid, leading to increased amount of rainfall, then the surface run-off and stream discharge will automatically increase which would cause local interruption in the cycle by accelerating the rate of erosion (rejuvenation), the effects of which may spread over larger areas.

Interruptions due to Base Level Change: Any change in the base level of erosion results into interruption in the cycle of erosion, or may initiate a new cycle. Changes in the base level of erosion are of two types: i) *positive change* due to rise in sea level, and ii) *negative change* due to fall in the sea level.

The base level changes caused due to sea level changes are eustatic (widespread). Base level also changes due to tectonic factors, e.g., subsidence of landmass causes positive change whereas upliftment causes negative change. Positive change of base level lengthens the cyclic time and accelerates the process of deposition. Contrary to this, the negative change causes accelerated erosion (rejuvenation). Rejuvenation, in fact, is the most important factor of interruption in the fluvial cycle of erosion.

The Rejuvenation: Rejuvenation means increase in the erosive power of the fluvial processes (river action) caused by numerous factors. Rejunevation takes place with the relative uplift with respect to sea level of a region which has developed a mature drainage system.

The rejuvenation may be caused from: i) a fall in base level, ii) an increase in stream discharge, and iii) a fall in sea level (eustasy) or an uplift of land (isostasy). Either of these processes will lead to new cycle of erosion. In general, rejuvenation will cause the river to incise itself into former valley, leading to the formation of incised meanders, knickpoints, paired river terraces, multi-storeyed valleys, and valleys in valley.

Important Features

Davis opines that in the development of land forms of a place, three factors play important roles. These factors are: i) structure, ii) process, and iii) time or stage. The factors have been called as the *Trio of Davis* which have been expressed as: "Landscape is a function of structure, process and stage (time)".

Structure: The word 'structure' has been used by Davis in a wider sense. It does not include only the regional geological structure (folds, faults, etc.) but also the nature of rocks and their physical and chemical properties, their permeability and solubility, the nature of joints, beddings and cleavage, etc. Thus, on a permeable rock, there will be minimum run-off and so erosion by running water will be slower than on an adjacent impermeable rock. Soft shales erode more readily than massive quartzite, while a rock of limestone or gypsum will undergo solution by underground water, in addition to being modified by any surface stream. As a rule, the influence of rocks on landscape is quite significant. Moreover, the structure, as a rule, is older than the landscape, and provides a base for the operation of the various gradational processes which give rise to new sequential land forms.

Process: The word 'process' includes the actions of all the endogenetic (internal) and exogenetic (external) forces. The endogenetic forces cause vulcanicity and diastrophism, and produce irregularities on the earth's surface by building mountains, plateaus, hills and undulating topography. On the other hand, there are exogenetic forces which tend to level down the earth's surface. The agents of these external processes of denudation are running water, ground water, glaciers, wind and sea waves. These agents erode the uplifted land, transport the eroded material and deposit them elsewhere.

Process is significant in the evolution of land forms. For example, alluvial fans, floodplains and deltas are the characteristics of fluvial process, while moraines, eskers and drumlins are found in the glaciated region. Davis considered running water to be the most universal process and initially based his cycle concept on fluvial erosion, but subsequently extended it to include other (glacial, wind, sea waves, etc.) processes also.

Time (Stage): In the evolution of land forms Davis has identified three stages. These stages of erosion are i) youth, ii) mature, and iii) old. Just as each process gives rise to distinctive landscape, so also each stage of the erosion cycle is characterised by distinctive land forms with the help of which it is possible to identify the stage of evolution of landscape. There is a direct relationship between the stage of development and the character of land forms. The stages of development of land forms, however, do not have equal life span.

The gorges, V-shaped valleys, waterfalls, rapids and steep spurs are the characteristics of youth stage. Meanders, ox-bow lakes, levees and floodplains are the features of mature stage, while monad nock, gentle slopes and featureless plains are the characteristics of old stage.

The main assumptions of Davis' model of 'geographical cycle of erosion' are:

1. Land forms are the products of the interactions of endogenetic (diastrophic) forces and the exogenetic (external) forces, originating from the atmosphere (weathering, running water, wind, ground water, sea waves and glaciers).
2. The evolution of land forms takes place in a sequential manner.
3. Streams erode their valleys rapidly downward until the graded condition is achieved.
4. There is short period of rapid rate of upliftment in land mass.
5. Erosion does not start until the upliftment is complete. In other words, upliftment and erosion do not proceed simultaneously. This assumption has been severely criticised by the subsequent geomorphologists.

Davis has prepared a graph to explain his model of geographic cycle.

The cycle of erosion commences with the upliftment of land mass. There is a rapid rate of short period upliftment of land mass of homogeneous structure. This phase of upliftment has not been included in the cyclic time as this phase is the preparatory stage of the cycle of erosion.

The main erosional and depositional processes and the resultant land forms of the youth, mature and old stages have been given briefly as under:

Youth Stage

1. There will be a few consequent trunk streams but few large tributaries. Numerous short tributaries and gullies will be extending themselves by head ward erosion and developing valley systems.
2. Valleys will have V-shaped cross profiles and will be shallow or deep depending upon the height of region above the sea level.
3. There will be a general lack of floodplain development except along trunk streams, and valley sides will rise from near the stream's edges.
4. Inter-stream tracts may be extensive and poorly drained. Lakes and swamps may exist in Inter-stream areas if these areas are not well above the local base level.
5. Waterfalls, cataracts and rapids may exist where stream courses cross beds of particularly resistant rock. They are most typical of early youth and will have disappeared before maturity is attained.
6. Stream-divides will be broad and poorly defined.
7. Valleys are V-shaped, characterised by convex valley side slopes. The overall valley-form is gorge or canyon.
8. Stream meandering may exist in youth.
9. The main river is graded.
10. The duration of youth stage is short.

Mature Stage

1. Valleys have extended themselves so that the region now has a well integrated drainage system.
2. Vertical erosion or valley deepening is remarkably reduced.
3. Adjustment of streams to such lithologic variations as exist may be evident in the existence of some longitudinal tributaries along the belts of particularly weak rock.

4. Stream-divides will be sharp and ridge-like resulting in a minimum of inter-stream uplands.
5. The lateral erosion leads to valley widening.
6. Lakes or waterfalls that existed in youth have been eliminated.
7. Floodplain tracts constitute a considerable portion of the valley floor.
8. Meanders may be conspicuous but, in contrast to those of youth, they are free to shift their positions over the floodplains.
9. The width of the valley floors does not greatly exceed the width of the meander belts.
10. The maximum possible relief exists.
11. The topography consists not so much of valley bottoms and uplands tracts as it does of slopes of hill slides and valley sides.
12. The marked reduction in valley deepening is due to substantial decrease in channel gradients, flow velocity and transporting capacity of river.
13. The duration of mature stage is longer to that of the youth stage but shorter to that of the old age.

Old Stage

1. The tributaries to trunk streams are usually fewer in number than in maturity but more numerous than in youth.
2. Almost total absence of valley incision. The lateral erosion and valley widening, however, continue.
3. Valleys are extremely broad and generally sloping both laterally and longitudinally.
4. There is marked development of floodplains over which streams flow in broadly meandering courses.
5. The valleys become almost flat, characterised with monadnocks, unakas, mosores and extensive undulating plain of extremely low relief.
6. Valley widths are considerably greater than the widths of the meander belts.

7. Inter-stream areas have been reduced in height and stream-divides are not so sharp as in maturity.
8. Lakes, swamps and marshes may be present but they are on the floodplains and not in inter-stream tracts as in youth.
9. Mass wasting and chemical denudation are dominant over fluvial processes.
10. Extensive areas are at or near the base level of erosion.
11. The duration of old stage is much longer to that of the combined duration of the youth and the mature stages.

The above outline attempts to present the regional picture and not that of individual valleys. A region may be classed as having mature topography and have within it individual valleys which are still youthful and others which may have attained early old age.

Positive Aspects of Davis Model

1. The 'geographical cycle of erosion' is the first scientific model, explaining the evolution of landforms.
2. It is a highly simple, lucid and applicable model.
3. The model is based on extensive and in-depth field observations.
4. It is a pioneer general theory of land forms development.
5. This model synthesised the current geomorphical concepts like 'base level', 'genetic classification of rivers', 'graded streams of G.K. Gilbert' and 'profile of equilibrium'
6. The model helps in describing, explaining, analysing and simplifying the evolution of land forms.
7. It also helps in improving the understanding of causal mechanism between the endogenetic and exogenetic forces and their resultant land forms.
8. One can make reliable predictions about the evolution of landforms with this model.
9. It is an analogous model in which landforms have been considered as living organism.

10. It helped in the formulation of several models, theories and laws about the evolution of Prefacelandforms and the processes of denudation.

The Evolution

The concept of 'geographical cycle' of Davis has been criticised on more than one count by the German and other geomorphologists. The main criticisms of Davis' model are as under:

1. Davis' assumption about the upliftment of landmass is not acceptable. He has described rapid rate of upliftment of short duration but as evidenced by plate tectonics it is an exceedingly slow and long continued process.
2. The processes of upliftment and erosion as postulated by Davis are erroneous. He opines that erosion does not start unless upliftment is complete. Can erosion wait for the completion of upliftment? The processes in nature are interdependent, and therefore, as the land rises, the processes of weathering and erosion begin. Davis answered this question. He admitted that he deliberately excluded erosion from this phase of upliftment because of two reasons: i) to simplify the model, and ii) erosion is insignificant during the phase of upliftment.
3. Davis' assumption about long period of crustal stability has also been challenged as the eventless long period is tectonically not possible. The plates are continuously in motion and the sea floor is spreading. The endogenetic forces always disturb the crustal stability.
4. Objections have also been raised about the 'time dependent series' of landforms as advocated by Davis. In the opinion of Albrecht Penck, 'geographical cycle' landforms do not experience progressive and sequential changes all through time. It was because of this point, he pleaded for deletion of time from the 'Davis Trio' of structure, process and stage (time).

Despite the given omissions and commissions, Davis' model of geographical cycle of erosion is a pioneer concept which has been adopted by most of the contemporary and subsequent geomorphologists all over the world. This model is the mother of all the models pertaining to the evolution of landforms. The concept of Davis with all its limitations and weaknesses is more significant and important and has acquired a special position in the study of geomorphology. The entire geomorphic thought has been influenced by the concept of Davis. The essential significance of his concept lies in the fact that it lays stress on the gradual evolution of landforms and provides a basis for their genetic classification.

Process of Erosion

Solution or Corrasion: Solution is the solvent action of the water as it flows over the rock. The degree of solution varies greatly as it depends on purity of water and solubility of rock.

Cavitation: Cavitation is caused by constriction of flow which raises the velocity, and hence, the kinetic energy of the stream. This, in turn, is compensated by a decrease in the pressure of water leading boulders bounced along during floods.

Hydraulic Action: Hydraulic action is performed by the lifting and quarrying effect of rushing water. The force of moving water dislodges rocks.

Impaction: Impaction is the effect of blow upon the river bed or banks by large.

Corrasion: The stream uses its load to scrape away its bed, particularly in steep confined sections of stream channels.

Attrition: Attrition involves the shattering and breaking up of the stream load through collisions and mutual abrasion leading to rounding of pebbles and boulders and decreasing their size.

These processes help the stream to:

1. Deepen its channel by down cutting of the stream bed.
2. Widen its channel by bank caving and undercutting.
3. Extend its channel by headward erosion or regressive erosion by streams and gullies.

Erosional Landform: V-shaped valleys, so called because of their resemblance to English letter V. These are produced by rapid cutting, and deepening so that stream floor remains comparatively narrow and the valley sides rise steeply giving the V-shape. The actual V-form is influenced by a number of factors:

1. Rock structure and resistance of the rocks to weathering and erosion, e.g., valleys cut in cohesive silts or clay have a tendency to create deeply cut ravines with steep slopes.
2. Aspect with shaded and sunny slopes experiencing different microclimates and different magnitudes in the operation of slope processes.
3. The rapidity with which river cuts, e.g., rapid period of uplift, will ensure that a narrow V-shape will be maintained.
4. *Climate:* This depends on whether the climate facilitates rapid mass movement or not. The three forms of V-shaped valleys are gorge, canyon and structural benches.
   (i) *Gorge:* A gorge has a steep precipitous wall within which a narrow river is confined. It can result from channel deepening as a result of recession of fall as has happened with Niagara gorge.

   Almost each continent has some examples of gorges. Thus the Indus, Sutlej, Brahmaputra and Arun form gorges in the Himalayan region where they are antecedent streams. Other examples are the Rhine gorge in Europe and the Zambesi gorge in Africa

   (ii) *Canyon:* A canyon has a remarkable similarity with gorge in that it also has steep slopes, narrow river bed and a predominant vertical erosion. The narrowness of the valley is determined by both vertical erosion and humidity of the area. The Grand Canyon, Colarado is very deep but its walls are not vertical because it passed through an arid zone where frost weathering and other forms have tended to open the V.

   (iii) *Structural Benches:* Sometimes hard and soft rocks may lie alternately in the V-shaped valley in a horizontal manner. Differential erosion leads to formation of step-like valleys which are known as *structural benches*.

Rapids: The part of the river where the current is flowing with more than normal swiftness is called a *rapid*. Rapids may be caused by a sudden steepening of the slope or by unequal resistance in the successive rocks traversed by the river. They are often due to outcrops of an unusually hard rock. At the outcrops erosion is less because of its resistant nature. The softer rocks are more easily eroded. Thus, the slope of the river above each outcrop decreases while below it the slope increases over the soft rocks, allowing the water to fall more rapidly. If the layer of hard rock dips gently downstream, rapids rather than waterfalls are formed. Many rapids have directly developed from pre-existing waterfalls by their recession Such rapids are common in the upper reaches of the river where structural control is important.

Waterfalls: If the face of the resistant rock becomes vertical, the stream plunges over the crest as a waterfall. Waterfalls can form due to many reasons, such as:

1. Due to the differential erosion of soft rocks when hard and soft rocks lie vertically.
2. Due to plateau scarp formation caused by upliftment and resultant steep precipitous slope, allowing the water to fall, e.g., Livingstone falls, Aughrabies falls, Gersoppa falls in India.
3. Due to creation of fault scarp such as Victoria falls on Zambezi.
4. Due to formation of hanging valleys when the tributary stream, which is at higher level joins the main stream.
5. Due to formation of glacial hanging valley, e.g., Yosemite.
6. Due to formation of knick point in a rejuvenating stream.

A fall that descends in a series of leaps is referred to as cascade. A cataract is a series of stepped water fall larger than cascade.

A cataract is formed when a river crosses a hard and soft rock bed. Cataracts are characterised by a number of chutes and rapidly descending fast flowing broken water which may take the form of rapids or merely two or three larger falls.

Pot Holes: Pot holes are cylindrical holes worn in the solid bedrock often at the foot of a rapid or waterfall. They are formed as a result of erosion, i.e., grinding action of the whirling particles on the bed rock caused by eddies formation. Rock fragments in the river load are carried sharply downwards by vertical turbulence adding to their impact velocity. They can also be swept backwards in the lee of some obstruction abrading to form a sharp line of separation between two smoothed surfaces. Horizontal turbulence will whirl particles of the river's load round in circles, particularly when they become trapped in crevices or joints in the rocks of the river bed. As the pebbles swirl round and round, they strike against the rock over and over again, gradually drilling a hole into the solid rock. These holes or potholes may enlarge and coalesce.

Transportation by Rivers: The effectiveness of stream as a geomorphic agent is its ability to move sediments and this ability to carry the sediment is dependent on the mechanisms involved.

Movement of Dissolved Load in Solution: Chemically dissolved material, for example, carbonates, sulphates, chlorides, and oxides are carried almost indefinitely by streams. Various amounts of organic matter are present and some streams are brown with organic acids. Velocity of the stream has negligible effect on materials that are dissolved. The deposition takes place only when the chemistry of the water is changed. Rivers carry dissolved loads of less than thousand parts per million but some streams, particularly those in arid regions carry several thousand parts per million. It has been found that when rainfall is abundant more than half of the material carried is in solution. In low relief areas, where the run-off is slow (in coastal regions), solution activity erodes the surface at a rate of about 50 mint per sq km per year or simply 1 m in 25,000 years. These dissolved loads are somewhat smaller in mountainous terrain eroding at a rate of 1 m in 1 lakh year to 3 lakh years. But, by no means, these rates are negligible.

Movement of Load by Suspension: The largest fraction of the material is moved by suspension. The suspended particles consist of mud, silt and fine sand and is seen in almost every part of the transporting segment of a river. The process of transportation is controlled by

1. Turbulence of the water, and
2. Terminal velocity of fall of each individual grain.

Terminal velocity is the constant rate of fall that a grain eventually attains when the acceleration caused by gravity is balanced by the resistance of the fluid through which the grain is falling. Terminal velocity increases with the size of the particle, if its general shape and density remains the same.

The movement of suspended particles is because the velocity of the upward currents in turbulent flow exceeds the velocity at which the sediment particles would sink in water. The bigger the particle, the more turbulent the flow is needed to keep it in suspension. And because turbulence increases when the velocity of the stream flow increases, it follows that the greatest amount of material is moved during flood time when velocities and turbulence are the highest.

Most silt and clay sized particles are distributed fairly evenly through the depth of the stream but coarser particles of same-size range are carried in greater amounts, lower down in the current, in the zone of greatest turbulence.

Movement of Intermediate Size Particles by Saltation (Latin: 'saltier '-to jump): Intermediate sized particles are too heavy to be in suspension and they bounce along the bed. Energy to leave the bed is supplied by hydrodynamic lift or by the impact of other jumping sand grains. Particles rise as much as 0.3 to 1 m and then travel downstream with the flat trajectory of a bullet. These are not distributed uniformly throughout the water like a suspension, they are basically restricted to the bottom of the stream.

Movement of Bedload by Traction: Some particles, such as coarse sand and pebbles, are too heavy for saltation. These particles called *bed load slide* or *roll along the bed* as a result of mechanical drop and push of turbulent waters. The bed load moves only when there is sufficient velocity to move the large particles, and thus, differs from suspended load, which moves constantly.

Bed load constitutes 50 per cent of the total load in some rivers but it normally ranges from 7 per cent to 10 per cent of the total sediment load. Bed load movement is a major tool of stream abrasion and they abrade both the sides and bottom of the stream channel.

The ability of a stream to transport is best defined by two concepts:

Capacity: It is the total potential weight of sediment load which streams can carry and varies with particle size.

Competence: The weight or size of the largest particles that the stream can possibly move along its bed.

The capacity and competence of a stream depends on:

Size of the Material: Since specific gravity of most surface rocks and minerals varies within narrow limits (2.5 to 3.5), particle size is primarily responsible for particle weight. In an area having uniform bed rock and an identical stream velocity, the distribution and proportion of the suspended and bed load are determined by their particle size and shape (although to a lesser extent).

Turbulent Velocity: Turbulent eddies maintain the upward force necessary to support the sediment in suspension while general turbulence will impart the energy to push and flip particles along the bed. Stream competence will be relatively great in shallow turbulent streams with a small hydraulic radius.

Current Velocity: The total possible amount of sediment of all sizes in transport will be largely determined by the volume of discharge which is greater in large streams than in small ones.

Process of Deposition by Rivers

Wherever the velocity of the river water is reduced, some of the load is deposited; a decrease in turbulence may also bring about the same effect. The heavier and coarser materials are first deposited followed by fine silt and mud. This material can be dropped at any point along a river's course. Deposition is not only confined to the lower course, it can even occur on the upper course, but here it is temporary and can be removed sooner or later.

Deposition Reasons

1. Break in slope: where the stream leaves the hills and enters a plain.
2. Widening of valley floor allows flooding to occur.
3. Contact with quiet water. When swift flowing stream enters quiet waters or lakes.

4. Where stream peters out as in arid and desert regions causing a decrease in river energy and allowing sediments to be laid.
5. Where rivers debouch into the sea.

Alluvial Fans and Cones: Where a heavily laden stream reaches the plain, its velocity is checked, it widens and much of its load is deposited. The deposited sediment spreads out as an alluvial fan. Deposition results from the sudden decrease in velocity as the stream emerges from the steep slopes of the upland and flows across the adjacent basin, with its gentle gradient. Where a river emerges from a mountain front into an adjacent basin its gradient is sharply reduced and deposition occurs. The channel soon becomes clogged with sediment and the stream is forced to seek a new course. In this way, the stream shifts from side to side as more of its energy is used in overcoming friction. With less energy for sediment transport, now it builds up arcuate fan-shaped deposits. Alluvial fans also develop where high competence tributaries converge with lower competence streams. Alluvial sediments sometimes accumulate to total thickness of over 300 m; the greatest value being attained a short distance below the apex of the fan. Normally, the coarsest sediments are found at the fanhead with finer away from it. But the debris are poorly sorted due to torrential deposition, especially near the fan head. The Kosi River in India has built a large fan.

An alluvial cone is a type of alluvial fan but one in which the slope angles are steeper (150 inclination) and the deposited material is generally coarser and thicker having been transported by ephemeral or short lived torrents, emerging from a high rocky massif at a mountain front or valley side. If conditions are favourable for evaporation and sinking of water, then the load is dropped gaining height and becoming an alluvial cone.

Piedmont Alluvial Plain: Where closely spaced streams discharge from a mountainous region across a piedmont, their deposits coalesce (i.e., merge laterally) at Alluvial fan the base of the mountain to form a piedmont alluvial plain. Examples of piedmont alluvial plain are the Indo-Gangetic plain on the southern piedmont of Himalayas, the Canterbury plains of New Zealand, the Argentine Pampas on the eastern side of the Andes Mountains and the American south west.

Flood Plain: A flood plain is an alluvial surface adjacent to a channel (on either side) that is frequently inundated. Flood plain size is generally related to the discharge of the river. The major ways in which floodplains may be formed are:

(i) By vertical accretion.

(ii) By lateral accretion.

(iii) By island formation and channel abandonment.

Vertical Accretion: When the stream is at the brink of overflowing (flooding) (when the river's channel is full with water to its brim) the state of flow of the river is known as *bankfull.* If the flood spills over, there is overbank flow. When large rivers, with gentle gradient and a predominantly suspended load, flood they spread out on both sides depositing a thin sheet of fine sediment material previously carried in suspension.

The force of water striking the stream bank causes erosion and undercutting which initiate a small bend in the river channel. With time, as the current continues to impinge on the outside of the channel, the bend grows larger and is accentuated leading to the growth of a small bend known as *meander bend.* On the side of the meanders, the velocity is minimum so that some of the sediment load is deposited. Since these deposits occur on the point of the meander bend, they are called *point bar deposits.* Erosion on the outside of the meander and deposition on the inside helps the meander to migrate laterally. As the meander moves sideways and downstream so do point bars building up an accumulation of sand and silt that covers the part of the valley floor over which the channel has migrated. Abandoned point bars within developing meander loops produce bars and troughs (swells). Lateral accretion of these point bars results in floodplain formation.

Lateral Accretion: All rivers naturally tend to flow in a sinuous pattern even if the slope is steep. This is because of the turbulent flow of water and irregularity or bend in the channel, which deflects the flow of water to the opposite bank.

Formation of Island and Channel Abandonment: Island formation and abandonment of channel are commonly associated with braiding (division of a stream into a complex pattern of several small channels divided by mid channel bars). Due to lack of competence and capacity, the load that the river carried is dumped in the mid channel to form islands. Since the rivers are constantly shifting, the side channels on the landward side become progressively filled with sediments until they are abandoned and the island was incorporated into the floodplain.

Flood plains are characterised by many distinctive features that include – levees, crevasse splays, backswamps, bayou, oxbows, scrolls, ridge and swale topography and yazoo rivers. Levees (French: *lever'-to* raise) are ridges lying parallel to the stream channel. In the case of large rivers they may rise more than 5 m above the general floodplain level. Levees are formed by rivers in times of flood when overbank flow causes a decrease in stream velocity after it has left the confining channel walls. Once this ability to transport its load has decreased (competence), much of the material is deposited, particularly the coarser particles along the edge of the channel. The formation of levees tends to raise the level of the river channel above the floodplain leading to widespread flooding, once the levees are overtopped or breached. When the breaching of levees takes place, water escapes through a series of distributary channels depositing coarse sediments called *crevasse splays.* The low lying flood plain adjoining a natural levee may contain marshy areas called *backswamps.* Having spilled over, the levees water may be diverted many kilometres downstream before it is able to regain the main channel. Re-entry will often be prevented by the presence of natural levees and the river continues flowing, parallel to the main river, following backswamp before a junction is possible again (deferred junction). Tributaries occupying such lateral positions have been termed yazoo streams, after the Yazoo River, a tributary of lower Mississippi River showing this type of down valley extension. It is implicit that in such cases the meander belt, of the major river does not occupy the full width of the floodplain rather it is bounded by sharply rising slopes called *bluffs.* With aggradation concentrated close to the meander belt, the position of the main or trunk river may ultimately become unstable. A lateral shift of the whole meander then occurs and the river comes to occupy one of its former flood basins. This process, known as *avulsion*, has characterised the growth of many floodplains of the world.

Within the confines of the meander belt, there is normally a down valley progression of individual channel loops. The meander, thus, may become constricted producing a meander neck. The meander neck is finally cut off as the stream cuts across the meander loop and follows a more direct slope. After some time, the meander loop is completely abandoned and forms a crescent-shaped lake called *oxbow lake* or *mort lake.* Gradual filling of the lake results in an oxbow swamp or bayous. It is characteristic of the lower Mississippi, especially in its delta.

Floodplains are of two basic types: a) Convex Floodplains, and b) Flat Floodplains.

Convex floodplains are convex in cross section, i.e., the land slopes away gently from the river bank to the sides of the valleys. The cross sectional convexity of a floodplain is primarily due to the nature and texture of sedimentation. The Tigris, Euphrates, Mississippi, Amazon, Mekong, Hwang Ho, Yangtze, Ganga and the Indus, all form convex floodplains.

Flat Floodplains: A large majority of riverplains are flat or gently concave in cross section. Natural levees are inconspicuous or absent and the alluvial flats rise very gently to the outer margins of the floodplain. Except, where small floodplains have experienced rapid sedimentation in response to accelerated soil erosion, most smaller streams have flat flood plains.

Flat and Convex Flood Plains

1. In flat flood plains, alluvium consists primarily of bed load where as in convex flood plains, the alluvium is basically flood silts and fine sediments.
2. Flat flood plains lack natural levees, back swamps and gathering streams.
3. The channels of flat flood plains are shallower and wider.

Delta Formation

Deltas are essentially formed where sediment laden rivers flow into standing bodies of water. This standing body can be a lake or an ocean. When a river enters a lake or the sea, its velocity rapidly diminishes and it deposits its load of sediment. The growth of a delta is complex phenomena, particularly in regions where large rivers deposit huge volumes of sediment. There are two major processes of delta formation – the splitting of a stream into a distributary channel system which extends into the open water in a branching pattern, and the development of local breaks, called *crevasses in natural levees*, through which sediment is directed and deposited in splays in between the distributaries.

As a river enters an ocean (or lake) and the flowing water is no longer confined to a channel, the current flows out and in the process loses velocity and flow energy. The coarse material carried by the stream is deposited in two specific areas (i) along the margin of quiet water on either side of the main channel (these build subaqueous natural levees), and (ii) in the channel at the river mouth where there is a sudden loss of velocity. Thus, they build a bar at the mouth of the channel. These deposits are able to create two small distributary channels, which extend seaward for some distance. As the process is repeated again and again, each new distributary is divided into two smaller distributaries leading to a system of branching distributaries, built seaward in a fan-shape pattern. Distributaries cannot extend indefinitely into the ocean because the gradient and capacity of the river gradually decreases. Therefore, the river is finally diverted to a new course, which has a higher gradient. This diversion takes place during floods when the river breaks through its natural levee and develops a new course to the ocean far away from the active distributaries of the delta. The shifting to a new course shifts the site of sedimentation to a different area and the abandoned segment of the delta is attacked by wave and current action. Thus, a new active delta is formed seaward. Finally, this may also be abandoned and another site of sedimentation may be formed. This shift in the channel helps in dispersing the sediment and the delta grows.

The manner in which the delta will grow will depend upon:

1. The relative densities of the river water and the standing water.

   Except where high concentrations of fine suspended sediments are carried by the debouching river, the dynamics of water mixing at the river mouth is strongly influenced by the relative densities of the inflowing and standing waters. There are three situations:

- Where the inflow is of equal density, it is termed homopycnal.
- Where the inflow is of great density, it is termed hyperpycnal.
- Where the inflow is of less density, it is termed hypopycnal.

Homopycnal flow naturally occurs where rivers flow into fresh water lakes and the simplest deltas develop where small rivers carrying a substantial portion of bedload flow into shallow lakes devoid of tidal or of significant wave forces. Here, the bedload reduces delta formation approximating a semi-circular form.

Hyperpycnal flow occurs when the river water is carrying large proportions of fine suspended sediments derived from the snout of glacier or where submarine slumps produce turbidity currents. Such flows spread graded beds of fine sediment far out into the basin of deposition.

Hypopycnal flow is the most common delta building accompaniment when fresh water rivers flow into salt water. Here, the coarse bedload is deposited as a shoal near the mouth and the coarse suspended sediment as levees flanking the currents into which the flow is subdivided by shoals. The finer suspended sediment is carried seaward in the surface plane jet which spreads out over the denser sea water and produces extensive gentle sloping submarine delta front. Such deltas build into deeper part in the absence of strong waves and tides, around an extending framework of coarser channel shoal and levee deposits cover offshore clays and inter distributary silts, clays and peats.

2. Magnitude and variation in the discharge of the river.
3. Amount and calibre of river load, particularly the proportion of bed load to suspended load.
4. Presence of a large river. This condition is necessary for marine deltas, otherwise the action of sea might disperse the sediment. In sheltered lakes, deltas may be built by smaller rivers.
5. Reasonably shallow water offshore so that the sediments can accumulate rather than disappear into a submarine canyon or an ocean trench. For example, the Congo, which virtually debauches into a submarine canyon, has no delta.
6. The geometry of the -coast, including its plan view and seaward slope.

7. The tectonic stability of the shoreline, particularly in the vertical sense.
8. The climate in so far as it exerts additional influences over such matters as the amount and type of vegetational cover and the growth of marine organisms.
9. A relatively sheltered coast, on which wave and current action is inhibited. The development of deltas is more likely in seas, such as the Gulf of Mexico, the Mediterranean, than on exposed coasts.
10. A small tidal range. The Mediterranean, Black Sea and the Caspian Sea bear witness to this in the Nile, Rhone, Po, Danube and Volga deltas. However, deltas can be built in areas of larger tidal range provided that offshore water is shallow and the coast has low wave energy. The Irawaddy, Ganga and the deltas formed by the Colorado (Gulf of California and River Fraser in British Columbia) are in such situation.
11. Flocculation caused by salt water. Flocculation is the process whereby clay particles in suspension coagulate on contact with seawater, become heavy and settle rapidly on the bed. All these factors, fluvial wave and tidal processes determine the morphology of the delta. But, this morphology is largely the result of complex interaction of the above variables.

Deltas can be classified according to their plan view geometry as either high destructive or high constructive deltas.

High destructive deltas are dominated by the removal of the river debris by wave and tidal currents such that distributaries do not become blocked and tend to remain stable in position. Wave influenced deltas are flattened in plan as there is rapid removal of debris by strong long-shore wave currents (e.g., the Sao Francisco River of Brazil).

Tide influenced deltas are arcuate and have funnel shaped distributaries which are kept open and straight by strong tidal scour and whose stability may be assisted in the tropics by the growth of mangrove vegetation, e.g., the Ganga, Niger and Mekong rivers.

High constructive deltas are dominated by large debris supply from the river.

Lobate (fan) deltas form where much of the river debris is coarse, giving a compact form over which the distributaries constantly change position. It is similar to alluvial fan, e.g., Nile and Rhone River deltas.

Bird's foot deltas occur when a river is delivering enormous quantity of fine sediment, and the formation of delta takes place around constantly shifting distributaries flanked by levees on which the effect of wave action is very meagre. Mississipi is the representative example.

The Purpose

Rejuvenation of rivers is caused by a wide range of factors. These basically fall into three categories:

Dynamic Rejuvenation

1. Upliftment of land.
2. Lowering of outlets.
3. Tilting of land.
4. Volcanic activity.

Static Rejuvenation

1. Climatic changes.
2. River capture.
3. Decrease in stream load.

Eustatic Rejuvenation

1. Locking of water in glaciers.
2. Changing capacity of ocean basins.

Dynamic Rejuvenation: Dynamic rejuvenation is brought about by various Earth movements.

Upliftment of Land: If the river is in its old stage or nearing old stage and the landmass starts uplifting, then the cycle is disturbed and the river is rejuvenated. Upliftment imparts a net input of higher potential energy availability leading to down cutting by the stream which now has a higher kinetic energy, and consequently, higher transporting power in comparison to load, and hence, greater energy availability for erosion and transportation.

Possible Changes in the Balance of Erosion and Deposition of a Drainage Basin

Factor	*Interfluves and Hillslopes*	*Geomorphic Response Upstream (Channel)*	*Downstream (Floodplain)*
A. In response to tectonic deformation of the drainage basin			
Increased relief and potential energy	Accelerated denudation	Accelerated bedrock dissection	Accelerated bed-load alluviation with increase in floodplain width and slope
Decreased relief and potential energy	Some reduction of denudation	Reduced dissection	Initial floodplain dissection followed by reduction of sediment calibre and floodplain width/slope
B. In response to changing sea level, affecting lower floodplain only			

Contd...

Factor	*Interfluves and Hillslopes*	*Geomorphic Response Upstream (Channel)*	*Downstream (Floodplain)*
Increased gradient, with sea level falling			Initial floodplain dissection followed by alluviation of suspended sediments on steeper, narrower floodplain
Decreased gradient with sea level rising			Accelerated alluviation of suspended sediments on a flood plain of reduced gradient
C. In responses to changes of vegetation and			

Contd...

Factor	*Interfluves and Hillslopes*	*Geomorphic Response Upstream (Channel)*	*Downstream (Floodplain)*
hydrologic balance on the interfluves			
Increased or accelerated run-off	Accelerated denudation with foot-slope colluviation possible	Alluviation	Accelerated bed-load alluviation on a wider, steeper floodplain; braiding
Decreased or decelerated run-off	Less denudation	Dissection of alluvium	Initial floodplain dissection followed by alluviation of suspended sediments on a narrower flood plain of reduced gradient

Lowering of Outlets: Outlets are the areas where rivers drain. If these outlets are lowered, then the velocity of the river increases, and consequently, it gets rejuvenated, resulting in a high degree of erosion leading to down cutting. If the streams are flowing into a lake, the water level of the lake is acting as the local base level. If the water level of the lake suddenly falls because of draining of lake water, then the rivers draining, into the lake suddenly descends abruptly into the lake. Thus, the rivers have been rejuvenated near their mouth. This leads to increased downcutting and consequently, youthful topography. This condition is likely to happen with Lake Erie when the Niagara will have drained its water. A condition when there is youthful topography near the lake and older topography higher up in the valley gives rise to a type of topographic unconformity.

River Capture: River capture is the action of a river acquiring the headstream of a second river by drainage area at the expense of the other. The process is carried out by a more powerful (captor) river which erodes its valley faster than its neighbour (the captured river). Steep gradients, higher volume of water, weak nature of the river bed relatively little load are some of the conditions that must be present which will allow one stream to capture another.

River capture takes place in three forms:

1. *Headward Erosion:* It does happen that streams extend themselves by headward erosion. In the process of extension, the streams cut back into their watershed until they both may join at the head. If this happens, then one of the stream is likely to take some water of the other stream. This will be determined by the depth of the channel, its velocity and volume of water. A river having greater channel depth, greater velocity and larger volume will draw some water from the other. This river is called the *captor river* or *captor stream*. The other river is the captured river, which after having captured becomes beheaded. The captured river has now a less volume of water commensurate to its size. This is called the *misfit stream* (small river in a big valley).
2. *Lateral Erosion:* When valleys of two consequent streams merge, one of the streams is captured, taking the joint flow of both of them.

3. *Intersection of Meanders:* In the old age when rivers are flowing in broad sweeping curves, two meanders may intersect, one of the rivers get captured acquiring a higher volume of water, and consequently greater erosive power.

Tilting of Lands: Tilting of land results in steepening of the stream gradient leading to increased down cutting by streams which now have transporting power in excess of that required for transportation of their load. The effect of seaward tilting, however, will be felt by those streams where courses run parallel to the direction of tilting and it is immediately reflected in deepening of the valley by the stream in it.

Volcanic Activity: Volcanic activity introduces local accidents such as the obstruction of the valley by lava flow. When a river in its old stage has been blocked by lava flow, it will try to remove this obstacle and in the process, there will be an intensification of erosive activity. If the whole landscape has been buried under lava flow, then a new cycle begins on the volcanic surface.

Static Rejuvenation: Static rejuvenation of the stream involves increase in its discharge which neither involves upliftment of the land nor eustatic lowering of sea level. This can happen because of:

Climatic Changes: Climatic changes, like increased precipitation in the river basin, will result in an increase in the volume of the river without any corresponding increase in load. There will, thus, be an increase in the river's energy, both to erode vertically and to transport the load.

River Capture: River capture increases the stream volume by diverting the drainage of one river (captured river) to another (the captor river).

Decrease in Stream Load: A stream in its headwaters starts by eroding vast quantities of unconsolidated glacial debris. As a result, such a stream is heavily loaded so that when it spreads out, deposition of sediments occurs. As time goes on and the supply of load down stream is lessened, the river flowing across the alluvial plain uses less of its energy in transporting its load, so that more energy is available for vertical erosion. The river is then rejuvenated.

Eustatic Rejuvenation: Eustatic rejuvenation results from causes that produce world wide lowerings of sea level rather than localised change in base level. Although, eustatic changes can be produced by a decrease in ocean basin capacity, associated with mid-oceanic ridge formation, it does not produce any substantial lowering of the sea level which is essential for rejuvenation of rivers on land. These changes are produced by glaciation.

Glacial Causes: Glacial causes are concerned with the amount of water locked up in ice sheets and glacier. Whenever there is a glacial period, sea water gets locked up in glaciers and the sea level falls. The fall in sea level during Pleistocene is a testimony to this. A fall in sea level has the same effect as an upliftment of land mass.

Peneplain, Lifted Up

A peneplain when it has been uplifted high above the present base level, so as to enter the second cycle of erosion, is an uplifted peneplain. The evidences of such uplifted peneplain are:

1. Accordance in interstream levels and summit areas.
2. Presence of topographic unconformities.
3. Truncation of rocks of varying resistance.
4. Presence of a thick zone of deeply weathered rock debris.
5. Presence of remnants of former alluvium.

Definition of Knick Points

A knick point is a break in the long profile of a river resulting from a fall in the base level. Because of the renewed down cutting a new long valley profile is created and where this intersects the long profiles of the former valley, a knick point will be found. Knick points are the first stage in the formation of river terraces.

The River Terraces: Just as a marine terrace represents an abandoned sea floor, so do river terraces represent valley floors abandoned by the rivers as they start cutting downwards to a new base level. A stream is graded as it reaches base level forming a wide flat valley. As the stream is rejuvenated, it cuts down through the first valley flat to a new base level where it develops a second valley flat inside and below the first one. Thus, spasmodic rejuvenation of the river takes place by uplift, so that it receives a series of impulses of renewed erosive activity leading to terrace formation. River terraces, formed due to fluctuations in sea level, are called *thallassostatic*.

Terraces may occur at an entire range of different heights, which may be either, of the same height on either side of the valley, in which case they are termed paired, or of different heights in which case they are said to be unpaired.

Where rejuvenation intervenes before the river has had sufficient time for lateral erosion to form a flat valley floor, there will be no river terraces at the side of the stream but merely breaks in slope in the valley sides. Such breaks mostly occur in deep mountain valleys where they show remarkable similarity with the topography produced by lithological differences or by glaciation.

The main characteristic of rejuvenated terraces is that they are essentially paired.

When there is a gradual lowering of a broad floodplain by a widely swinging meander belt, it produces terrace remnants which converge downstream, rather than being more or less parallel as in the case of paired terraces.

Stream terraces may also be controlled by the shape of the rock floor of the valley. They may also result from changes in climate. River terraces can be of various types and they reflect different geomorphological histories.

Spurs and Benches: Mature valleys are cut deeper after rejuvenation and lead to the formation of river terraces. If a tributary stream joins the main river, there will be formed a series of spurs and benches, the tops of which represent the old valley floor. Where there has been repeated spasmodic rejuvenation, there may be several benches at descending levels.

Incised Meanders: If a stream, that is meandering widely over its flood plain, is rejuvenated, it may proceed to cut a canyon or gorge into the old meanders because of the extra erosive capacity while still retaining the meandering form. These are called *incised meanders*. The best developed examples are those of the River Colorado, USA, where the uplift of 2200 m in the Tertiary period renewed downcutting to a fantastic depth. Incised ingrown or entrenched ingrown meander, some vertical erosion has cross section of the valley may be asymmetrical because of lateral erosion on the outside of the bend only. Thus, it has a steep slope on the outside of the bend and a gentler slip-off slope on the inside.

Entrenched meanders have been affected simply by vertical erosion and the valley cross section is even and symmetrical.

The differences between the two types are not perfect as there is every possibility of intervening phase depending on the rate of down cutting.

The floodplain characteristic of meander cutoffs, resulting in oxbow lake, is especially repeated in incised meanders, especially those of the ingrown type. Continued attack on the upstream sides of spurs and enlargement of the meander leads to a tendency for a break at the neck of the spur as it becomes thin. When the breach finally occurs, the old course of the meander is left as a curve through valley presumably drained by minor tributaries while the remains of the spur form an isolated meander core.

Under some very favourable circumstances, undercutting of the neck of the spur may lead to formation of a natural arch. This is caused by changes of form of entrenched meanders and the wearing back of the confining walls controlled by lateral undercutting of the river banks. As there is localised under cutting on both sides of the narrow neck of a constricted loop, a cave is excavated, particularly when the rocks at river level are weaker than those above. Finally, the two caves meet and the stream flows through the hole. The stronger rocks of the roof remain for some time as an arch spanning the stream and the loop shaped gorge is abandoned.

There are many examples of such type of arches in Utah, the most famous of which is the Rainbow Bridge.

Theory of Polycyclic Landforms

The concept of Polycyclicity is a by product of the idealistic "Davisian Geomorphic cycle" and its inadequacies to explain the complete process of landscape evolution through time.

Partial Cycles are far more likely to occur, for much of the Earth's crust is restive and subject to intermittent and differential uplift causing interruptions in the cycle. Nature or old age topography is likely to have superposed upon it youthful features as a result of Rejuvenation which is suggested by Topographic unconformities and may result from causes which are Dynamic, Eustatic or static in nature.

In nature, as is the complexity of landscape evolution more common than simplicity, so is the Multicyclic evolution more common than Monocyclic development.

Horberg presented a five food classification of landforms: .

1. Simple (One dominant process),
2. Compound (Two or more processes).
3. Monocyclic.
4. Polycyclic or Multicyclic.
5. Exhumed or Resurrected.

Monocyclic landscapes bear the imprint of only one cycle of erosion and are restricted to newly created land surfaces such as volcanic core or a lava plateau.

Polycyclic landscapes have been fashioned by two or more partial cycles. Much, if not most, of the world's topography is pòlycyclic and are found on all continents except Antarctica.

Measurement of landforms seldom give evidence of an ideal sequence of stages. In fact, what has actually happened is that the evolutionary process has been slowed down, interrupted brought to a stand still and sometimes even reversed. Any landform or tract of country, whose geomorphic features have been developed under several cycles or part cycles of erosion, which may be initiated by changes of base level or by climatic changes, is termed as polcyclic landform. Polycyclicity can result from changes in the climatic conditions. With accompanying variation in the dominant geomorphic processes and are alternatively referred to as Polyclimatic landscapes, e.g., the great Lakes of North America, Fjord coasts of Norway, etc. Different terrains exhibit elements from several different cycles not simply the current cycle. Thus, a region of stepped erosion surfaces with knick points on the rivers is an example of polycyclic landforms.

Polycyclic landscape can be simple or compound.

Other examples: The Rejuvenated landscapes, e.g., Paired river terraces, older alluvial plains, two cycles valleys, incised meanders are all Polycyclic in origin.

Landforms that are Fluvial

Despite the universality of wind, it is moving water, which plays the dominant role in the transport of weathering products and in the denudation of the Earth's surface. The first example of the action of moving water is provided by the rainwater, running down a slope. When this rain is torrential, tonnes of soil may be eroded. This erosion, called *sheet erosion* is unlikely to proceed in a uniform manner.

The sheet of rainwater is guided into depressions and areas unprotected by plants and the concentration of erosive action produces rills of irregular depths running down the sloped. A master rill finally emerges by flooding of water into them, and thus, an draining system is developed. The main streams flowing directly into the slope are joined by tributaries at accordant junctions, i.e., without a change in gradient. These streams which are relatively small are called *rivulet*. The master rivulet finally drains into a larger stream called *river*.

Rivers and their associated streams undertake two very important physical functions:

1. By draining the land surface they dispose the superfluous water brought by precipitation.
2. They are responsible for much of the denudation of the land surface over large parts of the Earth.
   (i) They dissolve and erode the rocks over which they flow.
   (ii) They transport the matter, which they have dissolved.
   (iii) They deposit some of the materials, which they have carried in suspension or roll along the stream bed.

As a result of all this, certain topographic features are produced which are called *fluvial topography*.

Rivers' Characteristic

1. Rivers are open systems in the sense that the mass of water and sediment entering is balanced by that leaving the system.

2. River acts as a system consisting of one main channel with all the tributaries flowing into it. A river system which lies in a drainage basin is bounded by a divide called *watershed*. Beyond this watershed the river is drained by another system. A typical river system has three subsystems:

   a. *Collecting System:* The collecting system consists of a vast network of tributaries in the head water region. These tributaries collect water and sediment and then funnel them into the main stream. The shape of the whole pattern resembles a tree, and hence, it is known as *dendritic pattern*.

      The collecting system has an initial network of tributaries and even the smallest tributaries have their own system of tributaries making the real number of streams phenomenal.

   b. *Transporting System:* The main trunk stream acts as channel ways for the movement of water and sediment from the collecting area to the ocean. While transportation is the major processes, deposition also occurs where meandering occurs, i.e., the river oscillates back and forth and also over flows its banks during floods. Thus, all the three processes—erosion, transportation and deposition occur in the transportation system of a river.

   c. *Dispersing System:* The dispersing system consists of many distributaries at the mouth of the river where sediments and water is dispersed into ocean, lake or dry basin. Deposition of coarse sediments and dispersal of fine grained sediments and water are the major processes involved.

3. There is a definite relationship between the tributaries and size of gradient of the stream. In other words, there is an order in the stream systems.

   a. The tributaries decrease in number in a mathematical progression downstream.

   b. The length of tributaries become progressively larger downstream.

c. The slope of tributaries increases exponentially downstream.

d. The channels in which the stream flows progressively become deeper and deeper down stream.

e. Although, the size of the valley is proportional to the size of stream, it increases down stream.

4. The flow in natural streams is complex and is influenced by various factors. The force moving a particle of water in the river is the component of gravity in the direction of flow. The magnitude of this component increases with the gradient of the river. However, frictional resistance by beds and walls limits the speed of the river. In a slow moving stream, the velocity increases steadily upwards from the river bed. There is a layer of motionless water on the river bed but above it, water moves smoothly in a laminae along straight paths parallel to the bed. This is called *laminar flow*. The laminar flow is disturbed particularly when irregularities, such as pebbles or boulders occur on the bed: eddies then move in various directions/and the flow is turbulent.

The flow is influenced by many factors:

Volume of Water or Discharge: Discharge=Velocity × Channel cross section area. Discharge increases downstream as more and more water is added from each tributary. Discharge can be enormous depending on the amount of precipitation in the river's collecting system, for example, the Amazon's heavy discharge is because of its location in the rain forest of the tropics of South America.

Velocity: The velocity of the stream channel is not uniform throughout the stream channel. Stream pattern, shape and roughness determine the velocity. Normally, the velocity is the greatest near the centre of the channel and above the deepest part (thalweg), where the frictional drag from the walls is minimum. When the channel is curved, as happens with meanders, the zone of maximum velocity shifts to the outside of the bend and the zone of minimum velocity is on the inside of the bend.

Velocity has a direct relationship with gradient — steeper the gradient, the more rapid is the flow, as it happens in the mountainous regions while low gradients produce slow sluggish flow. Finally, when the stream approaches the ocean, the velocity is reduced to zero. Volume also affects velocity greater the volume faster is the flow.

Shape and Size of the Channel: The shape of the stream channel affects the nature of flow and the velocity of the given volume of water along a particular slope. Deep narrow channels and wide shallow channels both have greater surface area per unit volume of water, and thus, the flow is reduced. The channels, which after least resistance to flow, have higher velocities; for example, the semi-flat channels cut their bottom and the channels become deeper. In either case, the tendency of the river is towards attaining an equilibrium between the flowing water and the surface area over which it moves making the channel shape approach towards semi-circle.

Slope of the Channel: The slope of the channel is steepest at the head decreasing downward. The slope of the channel is represented by a cross section of the stream from its head water to mouth galled the longitudinal profile.

Base Level: The base level is the lowest level to which the stream can erode its channel. This level is the sea level but a lake can also act as a temporary base level for the streams entering it. However, temporary base levels are only temporary and a lake forming local base level can be drained. Even sea level is liable to change, and consequently, the whole of the longitudinal profile of the river is changed.

Stream Load: The capacity of a stream to transport sediment increases to a third or fourth power of its velocity, i.e., if the velocity is doubled, the stream can move from 8 to 16 times as much sediment. The stream load (amount of material the stream transports at a given time) is usually less than its capacity (the total amount of sediment the stream can carry). The maximum particle size a stream can transport is its competence. The competence of some streams is enormous and they are able to transport boulders as heavy as 60-70 tonnes for several kilometres.

The flowing water from dam burst can transport boulders as heavy as 16,000 tonnes to the formation of air bubbles. As the stream channel widens and velocity decreases, the pressure is increased leading to collapse and implosion of air bubbles. This sends miniature shock waves, which have a strong effect on the nearby rock, and particularly on rock steps in the rapids and waterfalls. Continual repetition of the process finally breaks the rocks.

Concept of Fluvial Erosion: Erosion is the active wearing away of the stream channel wetted perimeter. The erosional work of running water acts in several different ways.

Reef Formation

Different Concept of Origin

Several theories and hypotheses have been propounded about the origin of barrier reefs and atolls by the oceanographers. Not even a single theory has, however, been considered adequate to account for explaining the varying conditions under which barrier reefs and atolls have formed. The main difference of opinion has been regarding the origin of the lagoons behind the barrier reefs, and the atolls. Some of the important theories about the origin of barrier reefs and atolls have been described as under:

Concept of Antecedent Platform

Rein (1870) and Murry (1880) independently suggested that still-standing submarine platforms could provide foundations upon which corals could establish themselves and develop into atolls or barrier reefs. About the genesis of lagoon, Murry explained that they come into existence as a result of the solvent action of water thrown over the reefs at high tide.

This theory has been criticised as most of the marine geologists do not consider solution by marine waters important enough to account for the formation of a lagoon. Moreover, most of the evidence points towards lagoons being areas of deposition rather than that of erosion.

Concept of Glacial Control

This theory was put forward by R.A. Daly. In his opinion, the atolls are the results of the influence of the changing sea levels of the Pleistocene period. In other words, the submergence was entirely post-glacial owing to the rise of sea level which resulted from the melting of Pleistocene ice sheets. The main points of Daly's theory are:

1. The depth of lagoons back of atolls and barrier reefs is remarkably uniform and rarely exceeds 80 to 90 metres, which implies a cause worldwide in nature.
2. Glacial conditions would result in a worldwide chilling of the seas and increased turbidity of oceanic waters as a result of the churning up of muds formerly below the reach of waves and thus would kill off reef-building organisms.
3. The destruction of these organisms permitted marine abrasion to attack the islands and banks upon which they had grown and produce at a lowered sea level a great number of truncated islands or benches around islands.
4. In the interglacial phase, the sea level rose. The relatively warmer and cleared waters favoured the re-establishment of corals and associated organisms upon submerged platforms which resulted in the development of an atoll or barrier reef. In such a situation, corals established themselves around the periphery of a platform and gradually drew upward and forward as sea level rose, until their bases were ultimately submerged about 250 to 300 feet. Daly's theory about the origin of the reefs and atolls has also been criticised by the geologists.

Some of the main criticisms are:

1. Many of the platforms are too broad to have been cut by marine abrasion during a glacial age.
2. Lagoon depths are hardly as uniform as have been claimed.
3. It is doubtful if the low-level marine abrasion could form the low-level platforms.
4. It is difficult to suppose that the low temperatures could be extended in the lower latitudes from the glaciated areas.

Principle of Subsidence

This theory was put forward by Charles Darwin in 1837 which was revised in 1842. Darwin postulated his theory after studying the Tahiti barrier reef and the Keeling atoll. His theory continues to be the most widely accepted theory about the origin of coral reefs.

Darwin observed that corals can live only in shallow water but the reefs formed by them going to great depth suggested subsidence as a possible explanation for his contradiction. In his opinion, the land or island along whose coast the coral reef started forming, has undergone a gradual subsidence, and as the rocky foundation of the reef subsided, the corals continued to build the reef upwards. In other words, there has been a broad correspondence or balance between the rate of subsidence and the rate of reef growth.

According to this theory, all reefs have started as fringing reefs and so long as there is no change in the relative level of land and sea, the reefs retain their fringing character. But, when the land or the ocean floor subsides and the fringing reef is submerged, the corals start growing upwards to keep themselves alive. Their rate of growth is more towards the outer edge. Consequently, it is only the outer margin of the reef which is able to keep pace with the rate of subsidence, and the side nearer the coast remains submerged, resulting in the formation of lagoon between the reef and the coast.

As the subsidence proceeds, the lagoon becomes bigger, and deeper and the fringing reef changes into a barrier reef. When there is further subsidence, and if the reef is situated around an island, the island is completely drowned giving rise to a central lagoon round which there develops a ring of reef or circular atoll.

Darwin's subsidence theory on the whole finds confirmation from recent seismic studies and drilling in the Pacific atolls.

The growth of coral reefs requires some favourable geographical conditions, as the coral polyps and the associated reef-building organisms require the following conditions for their successful growth:

1. The hermatypic corals and the associated organisms and algae which are most common reef builders are confined to the tropical belt. They require a temperature of 23° to 30°C. A winter temperature of less than 18°C and summer temperature of over 35°C are not conducive for their growth. The coral reefs are consequently confined to the warm tropical waters between 30°N and 25°S.
2. Corals can live only in saline water. The average salinity of water should be between 27 per cent to 40 per cent. The flood of fresh water kills the corals. Thus, corals are not found near the mouths of rivers and in areas of heavy rainfall.
3. The growth of corals is confined to shallow waters which are less than 65 metres (200 feet) in depth. It is because the sunlight does not penetrate beyond such depths and adequate oxygen is not available.
4. Corals require sediment-free clear water. The muddy water along the mouths of rivers is not favourable for the growth of corals.
5. Corals thrive well in the areas where ocean currents do not disturb the shallow sea waters.
6. Corals do not grow above the sea level, as they cannot survive for more than an hour or two above the sea level at the time of low tide.
7. Corals grow on submarine platforms which may act as foundations for corals. These platforms should not be deeper than 50 fathoms (about 92 metres).

Definition and Purpose

A reef created by the coral polyp along the shallow shores of some tropical seas is known as a *coral reef*. Actually, what is commonly called as *coral reef* was not built solely by corals. Numerous other organisms contribute to the growth of a reef, among them being numerous forms of *calcareous algae, stromatoporoids, gastropods, green algae, cyanbacteria, worms, oysters,* and *mollusca*. Moreover, a coral reef comprises a complex of several sub-classes of true coral growing *in situ*, together with debris of littoral shells, algal material and chemically precipitated carbonates of calcium and magnesium. The true corals may comprise only half of the total bulk of the limestone reef. Its growth, however, depends on the successful existence of coral polyp. A coral reef is also known as *bioherm* (organic mound). The coral reefs can be found in less than 2 per cent of the tropical ocean.

Although they look like flowers, corals are related to sea anemones and jellyfish. Some corals are solitary animals with bodies upto 30 cm (12 inches) in diameter, but most of them, more than 500 species, are ant-sized organisms crowded into colonies called *coral reefs*.

An individual coral animal, or polyp, feeds by capturing and eating plankton that drift within reach of its rosette of tentacles. Victims are entrapped by stinging cells on each tentacle, transported to central gastric cavity, and rapidly digested. Tropical corals feed at night, at dawn the polyps retract into their skeletal cups to withstand drying, should the colony be exposed to air at low tide. Coral polyps can also feed by direct absorption, a process in which they simply transport dissolved food molecules through their body walls. Tropical reef building corals are *hermatypic,* a term derived form the Greek word *hermatas,* which means mould builder. Their bodies contain masses of tiny dinoflagellates called *zooxanthellae (xanthos*=yellow). These single-celled plant-like organisms facilitate the rapid biochemical deposition of calcium carbonate into the coral skeleton.

Because of the needs of its zooxanthellae, hermatypic corals depend on light and warmth. Reef corals grow best in brightly lighted water about 5 to 10 metres (16 to 33 feet) deep. Coral reefs can form to depths of 90 metres (about 300 feet), but growth rates decline rapidly past the optimum 5 to 10 metre depth. In ideal conditions coral animals grow at a rate of about 1 cm (0.4 inch) per year. They prefer clear water because turbidity prevents light penetration, and suspended particles interfere with feeding. The animals are protected from the harmful effects of bright sunlight by a mucous coating that contains an ultraviolet blocking 'sun-tan lotion'.

Nearly all the reef-building corals are found within the 21°C (70°F) isotherm – a zone that corresponds roughly with the

25 degrees latitudes in both the hemispheres. Poleward of this area, water is too cold for the zooxanthellae to survive, temperature below 18°C (65°F) causes their death. Individual coral organisms are, however, found in some cold, high latitude waters as well. Such corals are found from Norway to the Cape Verde Islands and off New Zealand and Japan.

Different Kinds

In 1842, Charles Darwin classified tropical reef structures into three types: i) Fringing reefs, ii) Barrier reefs, and iii) Atolls.

Fringing Reefs: A coral reef, which is attached to the shore, either as a continuous wave-washed erosion platform or separated from the coastline by shallow lagoon. Beyond its seaward margin the ocean water deepens rapidly. The fringing reefs are generally not very wide, and where a river enters the sea from the land, they become broken and discontinuous. Such reefs are common along many tropical coasts.

Fringing reefs form in areas of low rainfall run-off primarily on the lee side (downwind side) of tropical islands. The greatest concentration of living material will be at the reefs' seaward edge, where plankton and clear water of normal salinity are dependably available. Most new islands anywhere in the tropics have fringing reefs as their first reef form. Permanent fringing reefs are common in Hawaiian Islands and in similar areas near the boundaries of the tropics.

Barrier Reef: This is an elongated accumulation of corals lying at low-tide level parallel to the coast, but separated from it by a wide and deep lagoon or strait. The deep lagoon does not permit coral growth. The lagoon may vary in width from a narrow channel to many miles.

They tend to occur at lower islands or in lines parallel to continental shores. The outer edge — the barrier — is raised because the seaward part of the reef is supplied with more food and is able to grow more rapidly than the shore side. The lagoon may be from a few metres to 60 metres (200 feet) deep, and it may separate the barrier from shore by only tens of metres or by 300 km (190 miles). In the case of northeastern Australia's Great

Barrier Reef, coral grows slowly within the lagoon because fewer nutrients are available and because sediments and fresh water run-off from the shore.

Great Barrier Reef: The Great Barrier Reef is the largest biological construction on the planet. It extends along the northeast coast of Queensland for 2,000 km (1,250 miles) and is upto 150 km (95 miles) wide. It is not a single reef, but a conglomeration of thousands of interlinked segments. The segments present a steep outer wall to the prevailing currents and trade winds. At a growth rate of 1 cm (0.5 inch) per year, the structure is obviously of great age. About 500 species of hermatypic coral live there, nearly 10 times the number found in western Atlantic Ocean.

Over 1,000 species of bivalue mollusks, 3,000 species of shore fish, and 40 species of sea snakes also live in the area. Between the main Reef and the mainland lies the lagoon — a shallow body of water, dotted with hundreds of islands. The reef was discovered by Capt. James Cook on June 11, 1770, when his ship, the Endeavour, rain aground on the reef that now bears the ship's name.

Coral Islands

Atoll is a ring-shaped island reef that encircles (sometimes completely surrounding) a central lagoon of sea water in which detrital material collects. It is composed of corals but in some oceanic atolls certain calcareous algae may form the bulk of the reef. It is particularly a common feature of the Pacific Ocean.

Coral debris may be driven onto the reef by waves and wind to form an emergent arc on which coconut, palms and other land plants take root. These plants stabilise the sand and lead to colonisation by birds and other species.

Some atolls are isolated, but most occur in loose groups in shallow continental shelf areas or in the deep open ocean. More than 300 atolls exist, mostly in the Pacific Ocean. They range in size from a few kilometres in diameter to Kwajalein in the Marshall Islands, whose slender 280 km (176 miles) ring of coral encloses a lagoon of 2,850 sq km (1,100 sq mi). The Lakshadweep (Laccadive) and Maldives Islands in the Indian Ocean are composed of atolls.

Winds' Role in Land Formation

'Aeolian' is a term pertaining to the wind; hence wind-borne, wind-blown or wind-deposited materials are often referred to as *aeolian landforms.* Wind is also an important agent of denudation. The geomorphic changes brought by wind are, however, less significant to that of running water. Wind action is quite pronounced in the arid and semi-arid areas of the world where the absence of vegetation cover and presence of extensive desolate rocks help in the erosional, transportation and depositional processes. Apart from the deserts, arid regions, wind action is also quite significant on sandy coasts, outwash plains of glaciated areas.

Effective Changes

Wind brings changes over the earth surface through the processes of: i) erosion, ii) transportation, and iii) deposition. A brief description of these processes has been given in the paragraphs that follow.

Erosion: Wind performs erosion in three different ways: i) attrition, ii) deflation, and iii) abrasion.

Attrition: The mechanism by which the particle size of any material is reduced by friction during transport by an agent of erosion (running water, wind, sea waves, etc.) is known as *attrition.* In the desert and semi-desert areas, there is rapid mechanical weathering and disintegration of rocks due to the great diurnal range of temperature and action of frost. When these rock particles are carried along by the wind, they strike against one another. Thus, by mutual friction and forceful contact they break down further which leads to gradual reduction in their size and ultimately they are converted into sand and fine dust particles. This process of gradual reduction in the size of rock particles by mutual impact is known as *attrition.*

Deflation: The complete blowing away of fine dust, leaving coarse and heavier materials is known as *deflation.* Through this process wind removes dry, unconsolidated sand, silt and clay from the land surface, especially in arid and semi-arid regions. The finer materials may be carried considerable distances before being deposited in the form of loess.

Deflation process may be observed in all types of regions (humid, semi-arid, and desert) but it becomes of outstanding significance in all arid regions. As a result of deflation are formed depressions or hollows known as *blow-outs* and bedrocks are exposed to wind abrasion (corrasion). Deflation is more pronounced in the areas of sandstone. Deflation has possibly been responsible for the formation of pedestal rocks which consist of residual masses of weak rock capped with harder rock.

There are numerous blow-outs (deflation hollows) in the valley of Nile. Of these the most well-known is the Qattara depression in the western part of Egypt. The Qattara depression has been excavated to nearly 135 metres below sea level. Similar but relatively shallower depressions are found in the deserts of Kalahari (Namibia), Simpson and Great Victoria of western Australia. Numerous very large depressions have also been found in the Gobi desert of Mongolia. The ground water table is the base level of wind erosion and limits the depth of deflation, though the water table may be lowered as the basins develop. Thus, the base level of wind erosion can be much lower than the ultimate base level of the sea.

Deflation commonly occurs in semi-arid regions where the protective cover of grass and shrubs has been removed by the activity of humans and animals. It does not occur where there are thick covers of vegetation or layers of gravel. The process is thus limited to areas such as deserts, beaches and barren fields.

Through the process of deflation wind removes mostly the fine particles (sands) and larger particles such as gravel are left over the surface. The accumulation of gravel over thousands of years forms *desert pavements* which protect the rocks below from further erosion.

Abrasion or Corrosion: In the process of abrasion, the winds pick up dust and sand and drive them with tremendous force against the rocks. In fact, in the desert and semi-desert areas winds carry with them enormous quantities of sand, dust and small angular fragments which act as tools of erosion as they strike against the rock surfaces. By this process, the less resistant rocks are etched and in time completely worn away, while the hard and very resistant rocks are polished and smoothed to a remarkable degree. Usually, the sand blast effect is greatest within 2 metres (about 6 feet). As the wind cannot lift sand more than a few metres, the effect of abrasion becomes insignificant after few metres in height. Through the process of abrasion, the most common landforms developed are mushroom-shaped blocks called *gara* in the Sahara Desert.

In areas where soft, poorly consolidated rock is exposed, wind erosion can be both spectacular and distinctive. Some pebbles, known as *ventifacts* (literally meaning 'wind-made'), are shaped and polished by the wind.

Erosional Landforms: Some of the important erosional landforms of wind action are briefly described below:

Zeugens (Rock Mushrooms): The tabular masses of more resistant rock resting on undercut pillars of softer material are known as *zeugens* or rock mushrooms. They are very often elongated in the direction of prevailing winds. Zeugens vary in height from less than a metre to about 30 metres.

Yardangs: Yardangs are steep-sided deeply undercut overhanging rock ridges separated from one another by long grooves or corridors as passageways cut in desert floors of relatively softer rocks.

Yardangs vary in size from a few metres to 1 km in length, upto 6 metres in height and 35 metres in width. They are separated from each other by a wind-scoured groove. They are elongated in the direction of the prevailing winds and are usually strongly undercut.

Dreikanter: This is a German term widely adopted to describe any boulder, cobble and pebble that has been shaped by aeolian sandblasting into a faceted rock. The faceted rock abraded for long period by wind erosion are called *vend/acts.* A ventifact rock boulder/cobble may have as many as eight abraded facets. The rock pieces (boulders, etc.) having three abraded facets are called *dreikanter,* while the boulders with two abraded facets are called *zweikanter.*

Wind Bridges: The continuous abrasion of a rock by powerful wind results into the formation of holes in the rock which are gradually widened. Such holes are called as *wind windows.* The holes are further widened and enlarged through the process of abrasion and deflation in such a way that an arch-like feature, having intact roof is formed. Such formations are called *window bridges.* Remarkable examples of these structures are found in the drier regions of western USA. The Rainbow Bridge of southern Utah is a magnificent arch of 95 metres (310 feet) with a span of 86 metres (280 feet).

Bowls and Caves: Depressions and caves are formed through the combined work of chemical weathering and wind. Usually they are formed in massive sandstones whose cementing material is calcium carbonate. This cement is easily dissolved by carbonic acid of the air or of groundwater. The sand grains thus freed are blown away by the wind. These bowls are also called as *animal traps.* It is nearly circular in shape with symmetrical overhanging sides. Such bowls may be seen in the desert of Turkmenistan, Egypt, Sudan, and Wyoming (USA).

Caves are formed in a manner similar to that described above. Caves of varying size, ranging from tiny, drill-like holes to those large enough to shelter many people are found particularly in massive sandstone and volcanic tuffs, on canyon walls and beneath the rim of buttes and mesas in many arid regions.

Inselbergs: 'Inselberg' is a German word (literally, island mountain) that has been widely adopted to describe a prominent steep-sided hill of solid rock, rising abruptly from a plain of low relief. It is a characteristic of tropical landscape, particularly in the Savanna zone, and is generally composed of a resistant rock, such as granite. Inselbergs in arid regions are also called as *bornhardts.* Passarge and Davis have taken inselbergs as the representative landforms of penultimate stage of arid cycle of erosion.

Demoiselles: Demoiselles is a term of French derivation used to describe an earth pillar (usually of *till)* in which a boulder has protected the underlying material from weathering and now survives as a cap stone precariously perched on the pinnacle of the slender pillar. The demoiselles are maintained so long as the resistant cap rocks are seated at the top of the pillars.

Desert Pavement: Where mountain-wash containing pebbles, gravels and sand is exposed to wind, the fine material is soon removed, leaving a mosaic of pebbles which has been fittingly called *desert pavement.* After long exposure to wind, the pebbles become highly polished. The pavement is common to deserts everywhere and is the most important factor in protecting the underlying fine-grained material from deflation. The degradation of desert surfaces is greatly reduced where the pavement exists.

Transportation: Wind transports dust and sand in several ways. The fine, light materials are picked up, suspended in the air and carried away. Heavy or large sand grains are dragged along the surface of the ground. In other words, wind transports sand and saltation surface creep. Silt and dust-size particles are carried in suspension. Transportation of eroded material by wind is not related to the slope of the land. Wind carries the load generally in suspended condition, and some particles are dragged along the surface.

The mechanism by which sediment is transported by bouncing or hopping along the surface is known as *saltation.* It is the most common form of wind transport in arid regions. The quantity of material transported by the wind depends upon the size and amount of particles and the velocity of wind. In general, the finer dust particles (silt and clay) are carried and deposited at long distances and the heavier materials are deposited at short distances.

Depositional Landforms: The main depositional landforms of wind are: ripple works, sand dunes, and loess.

Ripple Marks: Ripple marks are small scale depositional features of sand. This pattern is produced in unconsolidated sediments by the agents of erosion-like wind, sea waves and running water. Ripples may be either longitudinal or transverse.

Significance of Sand Dunes

A mound or ridge of wind-blown sand is known as a *dune* or *sand dune*. The dunes are generally mobile. There is wide range of variations in their shape, size and structure. They also vary in height, length and width.

On an average, their height ranges between a few metres to 20 metres, but there are some sand dunes which are more than several hundred metres in height and 5 to 6 km in length. The windward slope of a sand dune is generally gentle (5° to 15°) while the leeward slope is always relatively steep (20° to 30°).

The formation, height and length of sand dunes are directly related with the speed of the wind and the quantity of material it carries. If wind, blowing sand along just above ground, meets an obstruction, such as a fence-post, bush, or large rock, the force of the wind is checked and sand is deposited on the leeward side of the obstruction. The initial deposit of sand thus forms a further obstruction, causing more sand to be deposited. Soon a dune may be formed by the given process of deposition.

Sand dunes are the depositional features of the *sandy deserts*. Nearly 20 per cent of the surface of the deserts in the world is covered by sand. The deserts, in which sandstone is the predominant rock, are characterised by sand dunes. The deserts of Saudi Arabia, Sahara, Utah and Arizona (USA), Victoria and Simpson deserts of Australia, **and Gobi** (Mongolia) are the typical examples of sandy deserts.

Different Kinds

Sand dunes have been classified by the geomorphologists on the basis of morphology, structure, orientation, location and ground pattern. Bagnold (1953) recognised two basic types of dunes on the basis of forms: i) transverse, crescentic dunes (barchans), and ii) longitudinal dunes (seifs).

Classification of Crescentic Dunes

Class	*Type*	*Description*
Crescentic	Barchan	Crescent-shaped dune with horns pointed downwind. Winds are constant with little directional variability. Limited sand available. Only one slipface. Can be scattered over bare rock or desert pavement or commonly in dune fields.
	Transverse	Asymmetrical ridge, transverse to wind direction (right angle). Only one slipface. Results from relatively ineffective wind and abundant sand supply.
	Parabolic	Role of anchoring vegetation important. Open end faces upwind with U-shaped 'blow out' and arms anchored by vegetation. Multiple slipfaces, partially stabilised.
	Barchanoid Ridge	A wavy, asymmetrical dune ridge aligned transverse to effective winds. Formed from coalesced barchans; looks like connected crescents in rows with open areas between them.

Classification of Linear, Star and Other Dunes

Class	*Type*	*Description*
Linear	Longitudinal	Long, slightly sinuous, ridge-shaped dune, aligned parallel with the wind

Contd....

Class	*Type*	*Description*
		direction; two slipfaces. Can be 100 m high and 100 km long. The *'draas'* at the extreme is up to 400 m high. Results from strong effective winds varying in one direction.
	Seif	After Arabic word for 'sword'; a more sinuous crest and shorter than longitudinal dunes. Rounded towards upward direction and pointed downwards.
Star dune	Star	The giant of dunes. Pyramidal or star-shaped with three or more sinuous radiating arms extending outward from a central peak. Slipfaces in multiple directions. Results from effective winds shifting in all directions. Tends to form isolated mounds in high effective winds and connected sinuous arms in low effective winds.
Other	Dome	Circular or elliptical mound with no slipface. Can be modified into barchanoid torms.
	Reversing	Asymmetrical ridge form intermediate between star dune and trans verse dune. Wind variability can alter shape between forms.

Barchans (Barkhans) or Transverse Dunes: The barchan or transverse dunes are formed in the sandy deserts in which winds blow almost constantly from one direction. Barchans have their long axis at right angles to wind direction. They are associated with large quantities of sand and relatively weak winds and are not very common in the interior of great sandy deserts but form a long coastlines. The barchans and the transverse dunes have an asymmetrical profile with a gentle inclined windward slope and steep leeward slope. Since the sand tends to continuously fall from the crest to the leeward side, the leeward side is known as the *slip-face* or the *sand-fall slope.*

In a barchan, the convex gentle windward side extends laterally to the distal 'horns' or 'wings' which curve downwind on either side of the steeper concave slip-face of the leeward side.

The wind carries the blown sand to a maximum height of 30 metres up the gentler windward slope from which it slumps forward from the crest down the unstable slip-face, where the eddy motions help to scour the slope and maintain its concave shape. The wings develop because the rate of advance of the barchan is more rapid at the lower extremities than at the centre, i.e., the rate of advance of dune is inversely proportional to its height.

Although barchans may be found in isolation they usually occur in groups or belts in constantly shifting sand area (sand sea).

Barchan dunes usually rest on a flat, pebble-covered ground surface. The life of a barchan dune may begin as a sand drift in the lee of some obstacle, such as a small hill, rock, or clump of bushes. Once a sufficient mass of sand has formed, it begins to move downwind, taking on the crescent form. For this reason, the dunes are usually arranged in chains extending downwind from the sand source.

Where sand is so abundant that it completely covers the solid ground, dunes take the form of wave-like ridges separated by trough-like furrows. These dunes are called *transverse dunes,* because like ocean waves, their crests trend at right angles to the direction of the dominant wind.

The entire area may be called a *sand sea (erg)* because it resembles a storm-tossed sea suddenly frozen to immobility.

Longitudinal Dunes (Seif): Seif is an Arabic word, meaning sword dune and adopted to describe a knife-edged ridge of sand or longitudinal dune. Its axis lies parallel with the direction of the prevailing wind and may extend for many kilometres. Groups of *seif dunes* form long ridges with intervening wind-scoured troughs and dune chains often extend over 100 km in length and rise to heights of 200 metres. There is no agreement over their formation, some writers believing that they grow simply as a downwind extension of a sand-drift behind an obstacle, others opine that they represent the coalescence of the *'horn'* of partly destroyed barchans which have suffered a blow-out. Their spacing may be due to vortex flow.

A seif or longitudinal dune instead of being transverse to the prevailing wind is parallel to it. Many seifs in the Egyptian sand sea attain heights of 100 metres, and some in Iran are as much as 210 metres high. According to Bagnold, the width of a seif is roughly six times its height. Seif chains have been described which are 300 km long. Their crests form knife-like ridges with many peaks and sags. One side of the crest may be rounded, and the other may show collapsing front (slip-face) at right angles to the prevalent wind direction. There may be gaps or corridors between adjacent seifs in which bare desert floor is exposed. Contrary to the more commonly held idea that longitudinal dunes or seifs are built mainly by winds blowing sand rather constantly from one direction, Bangnold maintained that they grow in height and width largely through the action of cross winds and grow in length during periods when the prevailing wind is parallel to the trend of the seif chain. He believed that the seif is a modification of the fundamental barchan form produced by strong cross winds transverse to the prevailing wind direction, and that their knife-like crests represent barchan forms superposed upon a ridge of sand extending.

Sand Seas: The hot deserts in several continents are covered with wind-blown sand. These areas are referred to as *sand seas* or *ergs.* It has been calculated that 99.8 per cent of all wind-blown sand is in the great sand seas of the world. The largest are in Africa, Asia and Australia. In Africa, about 8,00,000 sq km (or one-ninth of the entire area of the Sahara) is covered by stable or active sand dunes. One-third of Saudi Arabia, approximately 1 million sq km, is covered by dunes and in the vast empty quarter, dunes may be over 200 metres high. The Australian sand seas are mainly in the western and central portions of the continent.

Star Dunes: A *star dune* is a mound of sand having a high central point, from which three or four arms or ridges radiate. This type of dune is typical of parts of North Africa and Saudi Arabia.

The internal structure of these dunes suggests that they were formed by wind blowing in three or more directions.

Whalebacks or Sand Levees: Sand levees (whalebacks) are flat-topped sand ridges which extend parallel to the prevailing winds but lack the collapsing fronts which mark seifs. They also have much larger dimensions. A whaleback may be 160 km (100 miles) long, 3 to 4 km wide, and 50 metres high. Locally, there may be seif forms upon their tops. They seem to be confined largely to the Egyptian sand sea.

Sand-sheet (Sand Drift): The term *'sand sheet'* is applied to a sand area marked by an extremely flat surface and absence of any topographic relief other than small ripples. The famous *Selima sand-sheet* of the Libyan desert is an outstanding example of sand-sheet. It covers an area of at least 3,000 sq mi and is perfectly flat. It consists of a few feet of sand resting upon bedrock.

Parabolic Dunes: The parabolic dunes were defined as long, scoop-shaped hollows, or parabolas, of sand with points tapering windward and with a windward slope much more gentle than the leeward, in contrast to the transverse dune. Parabolic dunes are generally associated with a cover of vegetation.

Migration of Dunes: Where they are not held in place by vegetation, dunes must invariably migrate in the direction of the prevailing wind. This is accomplished by the wind blowing sand grains up the gentle windward face of the dune and letting them fall down the steeper leeward face. In regions where the sand is very dry and the wind is very strong, migration is rapid. Even in humid regions forests and cultivated land have been completely overwhelmed by migrating dunes. In the drier parts, villages and cities have been buried. On an average, dunes migrate at the rate of few metres per year.

Loess: Loess is a wind-blown deposit of fine silt and dust. It is unstratified, non-indurated, calcareous, permeable, homogeneous and generally yellowish in colour. It consists of angular to sub-angular particles of quartz, feldspar, calcite, dolomite and other minerals held together with a montmorillonite binder. Unweathered loess is usually gray in colour but because of its permeability exposures of unweathered loess are uncommon.

The two important sources of loess are the weathered materials in deserts (hot loess) and the very fine powder from regions of glacial outwash (cold loess). The loess deposits are found away from the source regions and outside the deserts. The dust particles of loess are so small that they hold together even when dry and once deposited cannot be easily lifted by the wind. Loess is coherent (bound together) but not cemented and hence is permeable. Lack of stratification is a clear proof of its aeolian origin.

Extensive deposits of loess are found in many parts of the world. The most extensive deposit is to be found in the region covering much of the North European plain, the Asiatic steppes and the northern plain of China. Other important areas include North Africa, central USA and Argentina. The largest known loess deposit is in China. There, caves and houses are carved out of the thick deposits.

Fluvial Deserts' Landforms

In the deserts the rate of evaporation is higher than the rate of precipitation. In most of the deserts, the surface drainage is practically absent due to the scarcity of rainfall. In general, in the deserts, there are two types of streams: i) through streams which rise in humid regions and flow across the arid lands, and ii) temporary streams which rise in highlands within or on the borders of deserts, flow for short distances, only to lose themselves in the desert sands. There are, however, very few permanent streams in true deserts. The occasional rainfall in the deserts results into flash-floods. The short-lived torrents carry an immense load of solid matter, the product of desert weathering, and they sweep along so much material that they turn into mud-flows and soon come to rest. At other times the stream divides into channels, forming an alluvial cone or 'dry delta' usually at the mouth of the valley or at the foot of the hill slope where the gradient eases. This mass of loose unconsolidated material can then be attacked by the wind.

Some of the important landforms resulting from fluvial action in desert are: badland, pediments, bajada, and playas.

Badland: Badland is any landscape characterised by deep dissection, ravines, gullies, and sharp-edged ridges which have been created by fluvial erosion on rocks of relatively low resistance occurring in a semi-arid environment. The Chambal ravines present a typical example of badland.

Pediments: A gentle slope, cut in bedrock occurring below a markedly steeper slope and extending at a low gradient down towards a river or alluvial plain is known as *pediment*. The pediment is separated from the steeper upper slope by a relatively rapid change of slope angle in a transitional zone, termed the pediment zone. The pediment is generally, concave in profile.

Bahada (Bajada): 'Bahada' is a term derived from the Spanish language. It is used to describe the gentle, sloping surface leading down from a mountain front to inland basin in an arid or semi-arid region. It is composed of unconsolidated materials, such as sand, gravel and angular scree, which together mantle the underlying rock-cut (pediment).

Playas: 'Playa' is a Spanish term referring to a level or almost level area occupying the centre of an enclosed basin in which a temporary lake forms periodically. It is generally composed of stratified beds of clay or silt, deposited within the lake, that usually contain large amounts of soluble salts. The gentle slopes running down the lake are known as *bahada*. Lake Lap-Nor is a playa in the Tarim Basin (China). Playa lakes may last for days, weeks or even months before they are completely dried up by evaporation.

Different Kinds of Deserts

The above analysis of the desert landscape shows that they are of great variety. It is possible, however, to summarise six distinctive types of deserts as under:

Erg or Koum (Sandy or True Desert): The *erg* in the Sahara and Saudi Arabia, and *koum* in Turkmenistan are the true sandy deserts. They consists of vast, almost horizontal sand sheets or of regular dune lines, or of an undulating sand sea. It has been estimated that 25 per cent of the Sahara is covered with sand (17% dunes, 8% sand sheets).

Stony Desert: In a stony desert horizontal sheets of smoothly angular gravel cover the surface. This is known as the *reg* in Algeria and *serir* in Libya and Egypt.

Rock Desert (Hamada): Hamada is an erosional plain with a bare rock surface, wind swept and almost clear of sand. It may form a wind-smoothed pavement, or it may be diversified by *zeugen* and *yardangs.*

Desert Plateau: A desert plateau is crossed by canyon-like valleys of exotic rivers deriving their water from beyond the desert lands. It is characterised by isolated mesas and buttes and piedmont fans at the foot of steep slopes, as in Libya and western Egypt (Africa), and in Utah and Arizona (USA).

Rock Peaks Deserts: There are some deserts which are characterised by rock peaks, as Tibesti and Ahaggar of Central Sahara, the peaks of Sinai desert (Egypt-Asia), western Arabia and those of Baluchistan. With their harsh, serrated outlines, their steep, craggy faces cut into by *wadis,* rising from a swathing mantle of angular rock-waste, they form most prominent features.

Intermont Basins: Intermont deserts have inland drainage, characterised with rock pediments, alluvial spreads, salt marshes and fluctuating salt lake. Tarim basin, Tsaidan basin, and Dzungarian basin are some of the excellent examples of intermont basin deserts.

Cycle of Arid Erosion

A theoretical cycle of arid erosion was postulated by W.M. Davis. No special conditions, except aridity, are required at the beginning of a cycle. It may be assumed that the land has been uplifted by either folding or faulting. Mountain ranges of various heights are separated by valleys of varying depths. The bottom of some of the valleys may even be below the sea level. All the streams will be consequent. Each basin will receive the drainage of only the local streams. There is no thorough drainage from one basin to another. There are as many drainage systems as there are separate basins. The individual streams will vary greatly in length; many will disappear through evaporation or by sinking into the sandy slopes, some may reach the bottom of the valleys, each of which is the local base level for its own system of streams. It is understood that there is not enough rainfall to fill any of the valleys and cause an overflow.

Youth Stage: In the youth stage, the relief is the maximum, and as the cycle progresses there is progressive reduction in relief. There is disintegration of rocks by mechanical and chemical weathering and deposition of rock debris. At the occurrence of heavy rainfall, short, consequent streams cut ravines and V-shaped valleys in the mountain slopes. The debris is washed down the slopes to the valley floor, where valley filling begins. Alluvial fans appear at the mouth of the valleys. Seepage water collects in the bottoms of the valleys, forming playa lakes. The silt of streams is deposited in lakes. The lake waters become salty, due to the minerals carried in solution by the streams and by seepage waters.

In this stage deflation is active. The wind blows away the dry dust. Its sand blast effect aids in direct erosion. Dunes are formed here and there on the valley floors from the sand that has been washed down by the intermittent streams. The rivers follow different courses at different times, depending on the intensity of rainfall. The uplifted part is thus cut into several deep valleys which are unrelated to the structure and hardness of rocks.

The relief of the region is reduced. The stream-cut valleys expose bare rocks to the mechanical and chemical weathering; thus the rocks are prepared for removal by slopewash and streams during and following the next rainfall.

Mature Stage: On account of scanty rainfall, the progress from youth to maturity is slow. In mature stage, there is rapid reduction in relief. The mountain fronts are cut back by erosion, the divides are narrowed, and the intervening basins become wider and higher due to the valley filling. The receding mountain fronts may be separated from the alluvium in the valleys by narrow, rock-cut plains (pediments). The alluvial fans have coalesced to form broad alluvial fans *(bajadas)*. Deflation, however, may increase, due to the widening of the basins, which allows a broader sweep of the wind. Towards the end of maturity the playa lakes starts drying up and deflation hollows are formed.

Old Stage: In old age, the highlands will be largely worn away. Island-like hills *(inselberg)*, because of their superior resistance to weathering, will stand above the surrounding pediments of *bajadas*. The streams have nearly completed their work of reducing the land to the level of the lowest basin. Wind-scoured hollows and dunes become more numerous. If not held in place by salt crust, the silt of the playas has been largely blown away by the wind.

"At last, as the waste is more completely exported, the desert plain may be reduced to a lower level than that of the lowest initial basin, and then a rock-floor, thinly veneered with waste, unrelated to normal base level, will prevail throughout — except where monadnocks still survive" (Davis).

Wind erosion continues indefinitely, wearing the land lower until it may be brought to an elevation far below sea level. The only limiting factor, apparently, is the ground water table. Below permanent ground water, the rock and soil are moist. Wind cannot erode wet ground, therefore, *the water table is the base level for the arid cycle of erosion.*

Process of Desertification

The great deserts of the world were formed by natural processes over long periods of time, as continents migrated into dry climates produced in low-latitude, high-pressure zones. During development, a desert may expand and shrink in response to cyclic climatic fluctuations. The margins of deserts, therefore, have always been transitional or gradational to the adjacent, more humid environments.

The desert fringes have very delicately balanced ecosystems. Sparse vegetation serves to inhibit wind erosion, but when it is destroyed, the desert expands. Along the desert margins, human activity is commonly superimposed upon the natural processes that cause the expansion and contraction of the deserts. Grazing livestock and pounding of soil by hooves stress the ecosystem beyond its tolerance. This results in a degradation of the land and, in many cases, expansion of the desert. This process, in which productive land is converted into desert, is referred to as *desertification.* When the general climatic trend is towards increasing aridity, desertification can occur with remarkable speed.

Desertification does not occur in a broad, even swath that can easily be mapped along the desert fringe. Deserts advance erratically, forming patches on their borders, and areas far from the desert may quickly degrade into barren rock and sand. Desertification presents an enormous problem for human existence. About one-third of earth's land is arid or semi-arid, but only about half of this area is so dry that it cannot support human life. More than 600 million people live in the dry areas, and about 80 million live on land that is nearly useless because of desertification. The most severe problems are in Africa and Asia.

Definition of Desert

Deserts are arid lands where more water is lost through evaporation than is gained from precipitation. In other words, any area where the rate of evaporation is higher than the rate of precipitation is called a *desert*.

Several types of deserts are recognised in addition to the familiar hot deserts. They are: i) the dry deserts of rock and sand that is almost barren of plants, ii) semi-arid deserts of scattered trees, scrub, and grasses, iii) coastal deserts such as the Atacama of Chile, and iv) the deserts on the polar ice caps of Antarctica and Greenland.

Areas of Desert

As stated above, winds are more active in the sub-tropical arid and semi-arid areas which constitute about 34 per cent of the total land area of the world. In general, the regions that receive less than 25 cm (10 inches) of rainfall are deserts and the areas getting between 25 to 50 cm (10 to 20 inches) are semi-arid regions. The major deserts in which the erosional and depositional aeolian landforms may be observed are: i) Sahara (North Africa), Nubian Desert (Sudan), Rub-al-khali (Arabian Desert), Thar Desert (India and Pakistan), Dashte-Kavir, Dashte-Lut (Iran), Registan (Afghanistan), Kara-Kum (Turkmenistan), Kizil-Kum (Uzbekistan), Takla-Makan (China), Gobi-Desert (Mongolia); ii) Desert of Kalahari (Namibia Desert, Karroo Desert, Kalahari Desert); iii) South American Arid Zone (Desert of Atacama, and Patagonia); iv) North American Desert (Mojave Desert of California, and Arizona and Sonora Deserts of Mexico); and v) Great Victoria Desert, Simpson Desert, Gibson Desert and Great Sandy Desert (Australia). The deserts having mobile sands are called *ergs* (Arabic word, means moving sand). The largest *erg* (sandy desert) of the world in Rub-al-khali which stretches over 5,60,000 sq km.

Job of Wind

Wind is an effective geologic agent. It erodes and can transport loose, unconsolidated fragments of sand and dust. It transports sand by saltation and surface creep. Dust is transported in suspension, and it can remain high in the atmosphere for long periods.

The great deserts of the world form in low latitude regions in a zone roughly 30 degrees north and south of the equator. Wind is an effective geologic agent in these deserts, capable of lifting and transporting sand and dust, but its ability to erode solid rock is, however, limited. The main effects of wind, as geologic agent, are the transportation and deposition of sand and dust. At wind speed of 30 km/hour sand grains can be removed, but extremely rare gusts of 120 km/hour are needed to roll pebbles along the surface. Even the lightest winds are sufficient to lift dust into the air and to keep it in suspension for long periods of time. Abrasion by wind-transported sand thus aids in eroding and shaping some rock surfaces, but most are small-scale features. Most erosional landforms in deserts were produced by weathering, and by running water in times of wetter climate. They are in a sense 'fossil' landscapes, formed by processes that are no longer active. Even minor *'niches'* or *'wind caves'* and certain topographic features called *pedestal rocks,* which are often thought to be caused by wind erosion, are actually produced by differential weathering.

Although it is relatively insignificant as an erosional agent, wind is effective in transporting loose, unconsolidated sand, silt and dust. It is responsible for the formation of great 'seas of sand' in the Sahara, the Arabian and other deserts as well as the wind-blown dust (loess) covering millions of square kilometres in China, Central Asia, central USA and parts of eastern Europe. According to one estimate, about one-tenth of the earth surface is covered by wind-blown dust. Soils from these deposits constitute some of the earth's richest farmland.

Continued Flows

Water Resource Management

The earth has a finite amount of water, but fortunately the resource is renewable, and reusable. The fundamental objective of water resource management is to balance the water budget, ensuring adequate quality as well as quantity at a given time and place. Since natural climatic processes often do not create the desired balance locally, various techniques have been developed or proposed to control one or more phases of the hydrologic system.

The utilisation of surface water (runoff) and utilisation of underground water are quite old practises which predate written history. Damming and storage in wet area or during the rainy period for subsequent transfer to meet needs at other places or times has become a widespread practise for redistributing water used in irrigation or hydroelectric generation. Alteration of terrain features, soils and vegetation also affects run-off and storage. Applications of impervious coverings (for example, urban pavements) greatly increase run-off at the expense of soil moisture and percolation to groundwater.

Other approaches to water management entail modification of atmospheric processes. Suppression of evaporation from reservoirs can be achieved by covers, by vertical mixing to reduce surface temperature, or by windbreaks, whereas heating and mechanical stirring of overlying air accelerates evaporative loss. Thin chemical films (alkanols, for example), having a low vapour pressure have been used with moderate success to retard evaporation from relatively calm surface waters.

Changes in vegetative cover affect evapotranspiration. Cultivating practises such as weeding and mulching reduce soil moisture losses; clearing of forests usually results in decreased ultimately related to the ocean. Moreover, every aspect of the biosphere is interwoven with the hydrologic system and depends upon it for sustenance.

The hydrologic system is highly susceptible to human influence. We can build dams for water storage, flood control and irrigation, but this alters the system of groundwater, the balance of sediment input in a delta, and the concentration of salt in farmland.

The hydrologic cycle is a closed system with a definite budget of income, outgo and storage of water in each part of the system. For example, the amount of water that flows out of a river basin will equal the amount of water that falls from the atmosphere, less that lost by evaporation, transpiration and seepage into the ground. We can borrow one part of the system to enhance another, but there is always a 'pay-back time'. Irrigation illustrates this point: in the last 200 years, agricultural irrigation has increased at a phenomenal rate.

Indeed, at the present time, one-third of global harvest comes from 17 per cent of the world's irrigated cropland. Manipulation of the water budget in river systems at this scale is a significant interruption of the hydrologic system and exacts an ecologic toll. Each year some 3,000 cu km of water is spread out over croplands. This has profound effects on global water bodies as well as on the cropland receiving the water. The most significant effects are:

1. Waterlogged and salted cropland,
2. Declining and contaminated groundwater,
3. Shrinking lakes and inland sea,
4. Destruction of evapotranspiration and increased run-off.

Estimates of Available Global Water Supply

	Land area in thousands of sq km (sq m)		Mean annual discharge (water supply) in km³ per yr (BGD)	Mean annual runoff in mm (in.)		Population in millions/or 1995	2000
Africa	30,600	(11,800)	4,220 (3,060)	139	(5.5)	700	577
Asia	44,600	(17,200)	13,200 (9,540)	296	(12.0)	3400	3544
Australia-Oceania	8,420	(3,250)	1,960 (1,420)	245	(9.6)	28	30.4
Europe	9,770	(3,770)	3,150 (2,280)	323	(13.0)	728	733
North America (Canada, Mexico, US)	22,100	(8,510)	5,960 (4,310)	286	(11.0)	382	397
United States (48 states)	7,820	(3,020)	1,620 (1,231)	182	(7.1)	–	–
United States (50 states)	9,360	(3,620)	2,340 (1,700)	250	(9.9)	261	268
South America	17,800	(6,880)	10,400 (7,510)	583	(23.0)	311	360
Global (excluding Antarctica)	134,000	(51,600)	38,900 (28,100)	290	(11.0)	5610	6127

Distillation for the purpose of desalination or purification of polluted water pre-empts the normal atmospheric processes of evaporation, condensation and precipitation to provide water supplies of acceptable quality. Condensation in cooling towers releases latent heat, permitting reuse of water in thermal processes of factories and power plants.

An obvious way to offset a water deficit is to increase precipitation. Capture of fog drip on screens or filaments and collection of dew on cool surfaces are methods of enforcing precipitation to augment meagre fresh water supplies along desert coasts, but the amounts of water gained are small. Rainmaking by cloud seeding or other means cannot change the total global water supply, but it might accelerate the hydrologic engine to make more water available at a specific time and place. Schemes to enhance precipitation by inducing atmospheric instability or to steer rain-bearing storms by cloud seeding or other means are somewhat hypothetical at present.

Because the hydrologic cycle operates as a unified system, we cannot consider any part to be isolated and independent. The atmosphere is related to rivers and lakes, which are interrelated to the groundwater reservoir, and all are ultimately related to the ocean. Moreover, every aspect of the biosphere is interwoven with the hydrologic system and depends upon it for sustenance.

Examples of Methods and Proposals for Altering Natural Hydrologic Processes

---

Run-off and storage

Dams, weirs, levees, terraces
Ponds, reservoirs, tanks
Drainage, channel dredging
Irrigation, water spreading, flooding
Wells, groundwater recharge
Alter terrain, vegetation, soil
Impervious coverings
Induce melting

Evaporation and transpiration

Combustion, heating and cooling devices
Fans, windbreaks

Alter terrain, vegetation, soil

Reservoir coverings, surface film retardants

Vertical mixing of water bodies

Condensation and precipitation

Fog and dew capture

Industrial cooling towers

Distillation, desalination

Cloud modification

Induce instability of air

Steer storms

Growth in Agricultural Irrigation

Year	*Area Irrigated*
1800	20 million acres
1900	120 million acres
1950	230 million acres
1994	620 million acres
1999	690 million acres

The rivers of the world are becoming increasingly contaminated. For example, the Rhine River has been described as Europe's largest sewer, yet it supplies drinking water for more than 30 million people. But it is probably cleaner than the drinking water in some parts of the world where 17,000 children die from diarrhoeal diseases each day.

It is not only the river water of the hydrologic system that is dirty; the rains and snows of Europe, South America, North America, Russia, China, Indonesia, and India also carry pollutants. Chemicals and paniculate matter discharged into the atmosphere return to earth's surface as pollutants and acid rain.

In brief, the practical value of understanding the hydrologic system comes from the fact that it provides water on land for the many uses by humans and other life forms. A great deal of mental and physical effort is expended in the attempt to control it for our benefit. Conservation of water resources, indeed of most natural resources, depends on one or more of its phases. Where we cannot control it, we must adjust to its peregrinations.

Magnitude of Water Cycle

The hydrologic system is perhaps the most fundamental and significant hydrologic system operating on the earth surface. Its influence on the development of surface features is sometimes subtle, and sometimes dramatic. To conceive and understand the magnitude of the hydrologic cycle at the world level and its influence on the surface features is a difficult task.

The magnitude of the water cycle may be conceived by considering the volume of water involved. From measurements of rainfall and stream discharge, together with measurements of heat and energy transfers into the water bodies, experts and scientists have calculated that 4,00,000 cu km of water evaporates each year (or 1,100 cu km per day). Of this, about 3,36,000 cu km is taken from the oceans, and the remaining 63,000 cu km from the water bodies on the land. Likewise, of the total 4,00,000 cu km of water that evaporates annually, about 1,01,000 cu km falls as rain or snow on the continents. Most of the precipitation (from 60% to 80%) returns directly to the atmosphere by evaporation. About 38,000 cu km of water flows back to the oceans over the surface as rivers or beneath it as groundwater.

Thus, if the hydrologic system were interrupted and water did not return to the oceans by precipitation and by surface run-off from the continents, the sea level would drop one metre per year. All of the ocean basins would be completely dry within 4,000 years. The recent glacial epoch demonstrates this point clearly that the hydrologic system was partly interrupted because much of the water that fell on the Northern Hemisphere froze and accumulated to form huge continental glaciers. This prevented the water from flowing immediately back to the oceans as surface run-off. Consequently, the sea level dropped more than 100 metres during the ice age.

As a final observation, consider the volume of water running off earth's surface. The circulation of water in the hydrologic system provides a continuous supply of surface run-off over the land. At any given instance, about 1,260 cu km of water are flowing in the world's rivers. Gauging stations on the world's major rivers indicate that if the ocean basins were empty, run-off from the continents would fill them completely in 40,000 years (This estimate assumes no precipitation falling directly into the oceans). Moreover, if all of the evaporated water returns to the oceans by run-off, with no return to the atmosphere by immediate evaporation and transpiration (return of water to the atmosphere by plants), then the ocean basins would be filled in slightly more than 3,000 years.

Gravity plays an important role in the hydrologic system. It is the force that causes rain to fall on the land and to move downslope. It causes rivers to flow and to transport sediments from higher to lower elevations. Indeed, the movement of water in each part of the hydrologic cycle is a type of gravity flow system. The hydrologic system eroded the Grand Canyon, carved the fantastic Himalayan mountains, and deposited the Sunderban delta, Nile delta and Mississippi delta. It formed the glaciers of the last ice age, and developed the Sahara desert. There is little of earth's surface that is not affected in some way by the work of the hydrologic system.

Global Oceans and Seas: The ocean, earth's greatest repository of water and one of the last great scientific frontiers, has been of great interest to geographers. Oceans and seas are important for food, minerals, transportation, and also for climatic control through interactions with the atmosphere. From the geographical point of view, oceans and land surfaces are not evenly distributed.

Systems of Glaciers

In cold climates, precipitation falls in the form of snow, which remains frozen and does not return immediately to the ocean as surface run-off. If more snow falls each year than melts during the summer months, huge bodies of ice build up to form glaciers. Large valley glaciers originate from snowfall in the high country and slowly flow down valleys as rivers of ice. They melt at their lower ends and return their water to the hydrologic system as surface run-off.

At the present time, the continent of Antarctica is covered with a continental glacier, a sheet of ice from 2 to 2.5 km thick. It covers an area of 13 million sq km — an area larger than the United States and Mexico.

The formation of a continental glacier completely modifies the normal hydrologic system because water does not return immediately to the ocean as surface run-off but moves slowly as flowing ice. An ice sheet similar to that now on Antarctica covered a large part of North America and Europe during the last ice age and retreated only within the last 15,000 years. As the ice moved, it modified the landscape by creating numerous lakes and other landforms.

Groundwater Systems: Another segment of the hydrologic system is the water that seeps into the ground and moves slowly through the pore spaces in the soil and rocks. As the groundwater moves, it dissolves soluble rocks and creates caverns and caves, which can enlarge and collapse to form surface depressions called *sinkholes*.

Problem of Pollution

When surface water is polluted, groundwater also becomes contaminated because it is recharged from surface water supplies. Groundwater migrates slowly compared to surface water. Surface water flows rapidly and flushes pollution downstream, but sluggish groundwater, once contaminated, remains polluted virtually forever.

Pollution can enter groundwater from industrial waste, injection wells, septic tank outflows, seepage from hazardous waste disposal sites, industrial toxic waste dumps, residues of agricultural pesticides, herbicides, fertilizers, and residential and urban wastes in landfills.

Systems of Shoreline

The hydrologic system also operates along the shores of all continents, islands, and inland lakes by the unceasing work of waves. The oceans and lakes are bodies of mobile water subject to a variety of movements — waves, tides and currents — all capable of eroding the coast and transporting vast quantities of sediments. The effects of shoreline processes are seen in wave-cut cliffs, shoreline terraces, deltas, beaches, bars and lagoons. Suspended mud is carried beyond the shore by currents and is eventually deposited on the sea floor.

The hydrologic system expends a considerable amount of energy upon the shore. Over a period of only a few tens of years, significant modification in the shape and configuration of many coasts can be seen. Hurricanes can greatly alter the sandy coastline in a single day.

Role of Aeolian Systems

The hydrologic system also operates in the arid regions of the world. In many deserts, river valleys are still the dominant landform. There is no completely dry place on earth. Even in the most arid regions, some rainfalls occurs, and climatic patterns change over the years. River valleys can be obliterated, however, by the great seas of windblown sand that cover the desert landscape. Air circulating in the atmosphere constitutes *aeolian systems,* which can transport enormous quantities of loose sand and dust, leaving a distinctive record of the operation of the wind. In the broadest sense, the wind itself is part of the hydrologic system, a moving fluid on the planet's surface.

Distribution of Streams

Streams contain a tiny fraction of all water and the smallest volume of any of the fresh water categories. Yet streams are the portion of hydrologic cycle on which we most depend, representing four-fifths of all the water we use.

A stream's flow rate is termed its discharge, usually expressed as the volume of water that passes a given point per unit of time. In volume of run-off, the Western Hemisphere exceeds the Eastern Hemisphere, primarily because the Amazon River carries a far greater volume than any other river, and because four of the world's highest discharge river systems are in the Western Hemisphere (Mississippi-Missouri, Orinoco, St. Lawrance, and Mackenzie). The Atlantic Ocean receives about 1.5 times more run-off than the Pacific Ocean.

Cycle of Hydrology

Hydrologic cycle is a simplified model of the continuous flow of all forms of water on, in and above the earth's surface. It has been defined as total plan of movement, exchange, and storage of earth's water in gaseous state, liquid state and solid state. In other words, hydrologic system is the complex cycle through which water moves from the ocean to the atmosphere, to the land and back to the ocean again. It involves water in rivers, lakes, glaciers, and oceans, in the atmosphere, and in the pore spaces in the rocks beneath the surface.

Of all the resources of the earth, none is more fundamental to life than water. The properties of water in its three physical states (liquid, solid and vapour) make it by far the most useful of compounds. We can breath it (water vapour), drink it, bathe in it, travel on it, or see beauty in its different forms. It is a raw material, source of power, waste disposal agent, solvent, medium for heat transfer, or coolant as the needs of modern technology may require.

The high specific heat of water, its ability to exist in gaseous, liquid or solid forms under natural conditions, and its capacity for storing or releasing latent heat with changes of state give it immense influence on atmospheric processes. At the same time, making it available at different times and places is a function of weather and climate.

The restless atmosphere is the most active agent in the constant redistribution of water on the earth's surface — a fact that becomes even more striking when we realise that only a minute fraction of one per cent (0.0001%) of earth's water is contained in the atmosphere, and if, at any time, all the atmospheric moisture were precipitated, it would create a layer averaging about 2 mm deep over the entire globe.

The seas and oceans contain about 93 per cent of the earth's water, 2 per cent is in the form of ice caps and glaciers, and fresh water bodies, ground water, soil moisture and vegetation account for about 5 per cent.

World Water Resources

	Area covered (million sq km)	*Volume (million cu km)*	*Per cent of total volume*
World oceans	360	1370	93
Ice caps	16	24	2
Terrestrial water.	134	64	5
Atmosphere	–	0.013	0.0001
Total	510	-1500	100

Past Theories

Hydrologic cycle is not a revolutionary but an evolutionary concept. It gradually evolved over many years, and is based on the collective observations of many scholars.

Throughout the history humans have wondered about the origin of rain, the source of rivers and springs, and the interactions of these waters. At first it was believed that rainfall could not account for the amount of water carried by rivers. Part of the reason for this was that early philosophers lived in the Mediterranean area, where the headwaters of the major rivers, such as the Nile, were far away. The idea of completed cycle in the movement of water from the oceans, to the rivers, and back to the sea again was recorded in the religious scriptures like Vedas, Bible, and Holy Quran. For example: "All of the rivers run to the sea, yet the sea is not full; unto the place from whence the rivers came thither they return again".

Regardless of this ancient observation, prior to the 16th century it was still generally believed that water in rivers and springs could not be derived from rain because it was thought that rainfall occurred in insufficient amounts. Leonardo da Vinci (1452-1591), an exceptional genius in many fields, is generally credited with one of the earliest accurate descriptions of hydrologic cycle. He writes: "Whence we may conclude that the water goes from the rivers to the sea and from the sea to the rivers, thus constantly circulating and returning, and that all the seas and rivers have passed through the mouth of the Nile an infinite number of times.... The conclusion is that the saltiness of the sea must proceed from the many springs of water which, as they penetrate the earth, find mines of salt, and these they dissolve in part and carry with them to the ocean and other seas, whence the clouds, the begetters of rivers, never carry it up."

But such a cycle was not demonstrated as fact until the mid-17th century. The discovery was made by two French scientists — Pierre Perrault (1608-1680) and Edme Mariotte (1620-1684). They independently measured precipitation in the drainage basin of the Seine River and then measured the discharge into the ocean during a given interval of time. Their measurements proved that precipitation alone produces not only enough water for the river to flow but also for springs. Precipitation, thus, was recognised as the basic source of all surface water. In addition, Edmund Halley (1656-1742), the English scientist after whom Halley's comet was named, estimated the evaporation from the Mediterranean Sea. He then concluded that it was as great as the volume of water in all the rivers flowing into the Mediterranean. These observations provided the initial basis for our understanding of the hydrologic system.

During the 16th and 17th centuries, there were many heated debates between natural scientists and Christian theologians, and new ideas propounded about nature were carefully scrutinised by the church. Although initially rejected by theologians, the hydrologic cycle gained favour when its usefulness in support of the doctrine of the *'Divine Plan of Nature'* became clear. This doctrine held that God created earth expressly for man and that everything within it was part of a *'Great Divine Order'*. Early versions of the hydrologic cycle were seen as scientific verification of the Divine Plan by science. An idealised model of the hydrologic cycle was presented by John Ray, a 17th century theologian and soon became generally accepted. Over the succeeding centuries, a more or less standard model has evolved. This model appears in academic textbooks today in a form very similar to that presented in the 17th century.

Distribution of Water in Hydrologic System: The total water of the planet is distributed in various sub-systems — in the air, on the land, and in the oceans — as well as in three states, viz., gas, liquid and solid. Compared to the total mass of the earth, the mass of water in the hydrosphere is extremely small, i.e., only one part in 4,500. Yet water covers 71 per cent of the surface of the earth and is extremely mobile, constantly moving from one place to another, often at remarkable speed. The major reservoirs are the oceans and the glaciers, which contain 99.5 per cent of the total water of the world. It may also be noted that more than 97 per cent of the water on earth's surface is in the oceans. Glaciers contain about 1.9 per cent, ground water 0.5 per cent, rivers and lakes 0.02 per cent and the atmosphere 0.0001 per cent.

Oceans: About 97.2 per cent of the combined liquid water, ice, and water vapour is in the vast reservoir of the oceans. There is only one output of any importance from the oceans, i.e., evaporation, whereas of the two inputs, precipitation accounts for about 90 per cent and run-off from the land about 10 per cent. The residence time of water in the oceans is about three thousand six hundred years. This, of course, is only an average time because only the water at the surface of the ocean is, involved in evaporation. Thus, water near the surface may have a much shorter residence time than deep water, which may be stored for very long periods, possibly hundreds of thousands of years.

Ice: Water in the form of ice constitutes about 80 per cent of the water that is not in the oceans, or about 2 per cent of earth's total. Most of this form of water is located in the great glaciers that cover

Greenland and Antarctica. Ice within the glaciers moves slowly from the areas where it accumulates to the margins of the glaciers, where it melts. Water may reside in the glaciers for thousands or even millions of years. Present estimates, based on rates of loss by melting, suggest that water resides in glaciers, on the average, for 10,000 years.

The amounts of water stored in the form of ice and in the oceans are intimately related. When glacial ice increases significantly, the sea level drops. And, when glacial ice decreases, the sea level rises. If the existing glaciers melted completely, the volume of water in the oceans would increase by about 2 per cent and the sea level would rise about 100 metres.

Groundwater: The next largest water reservoir is groundwater, i.e., the water contained in the pore spaces of rocks and soil. Surprisingly, about 20 per cent of the water not in the oceans occurs as groundwater.

Soil moisture accounts for about 0.005 per cent of earth's water, or about 0.5 per cent of the water not contained in the oceans. Even at this low percentage, there is more water in the soil than is contained in river channels. There are two components of soil moisture: i) water that is migrating down to the water table, and ii) water that is retained in the soil. Water in the soil is withdrawn by evaporation from the soil surface and by plants. Residence time for water in the soil is about one month. Soil moisture is important because it migrates into rivers and groundwater systems, but it is also of supreme biological importance because it is the essential foundation for plant and animal life.

Lakes: Lakes contain approximately 0.017 per cent of earth's water, or about 0.7 per cent of that not contained in the ocean basins. Slightly more than half is held in fresh water lakes (lakes with outlets); the rest is in saline lakes (lakes without an outlet). About 75 per cent of the total volume of fresh water lakes occurs mainly in the large lakes of North America and East Africa, and Lake Baikal in Russia. The Caspian Sea contains 75 per cent of the water in closed basins with no outlet to the oceans. Water resides in the Caspian Sea for about 200 years. This relatively short residence time is due to the high rate of evaporation, which is equivalent to the discharge of a large river. Residence time for Lake Superior is almost 200 years whereas for Lake Erie, 90 years. These figures are alarming in the sense that serious pollution of lakes can be difficult to correct in bodies where water resides for a considerable length of time.

Atmosphere: The amount of water in the atmosphere is surprisingly small (0.0001%) when compared to that in other parts of the hydrologic system. Measurements show that if all the water held in the atmosphere condensed at a given moment and rained evenly over all of earth, it would form a layer only 2 mm deep.

The volume would be roughly equivalent to the volume of water carried by the world's rivers. The daily rate of exchange between the water in the atmosphere and that on the earth includes approximately 2.5 mm that falls as rain and 2.5 mm that evaporates from the surface. Calculations show that the average residence time of water in the atmosphere is about 10 days, and there is a complete exchange of atmospheric moisture forty times a year.

This extremely short residence time underlines the important fact that the atmosphere is very dynamic and a most vital link to other parts of the hydrologic system. The dynamic aspect of atmospheric water can be appreciated by watching time-lapse motion pictures of the regional movement of clouds on weather programmes. During a relatively short time (a few hours), the clouds may appear to explode into a storm and then disappear as they move over the surface.

Rivers: At first glance, a river may seem to be a primary reservoir of water in the hydrologic system, but actually the volume of water in earth's river systems is only about 0.0001 per cent of the total water on earth, or 0.005 per cent of the water not in the oceans. The volume of water in river systems at any given time is not a measure of its importance. Water flows through rivers at an average rate of 3 cubic metres per second. At this rate, a given volume of water could traverse the entire length of the longest river in only 20 days. This means that although the volume of water in rivers at a given time is small, the total volume going through them over a period of time can be enormous. As water moves, it erodes and transports huge volumes of rock and soil debris. No other system is so universally important in shaping the earth's surface as running water.

The input of water into river systems is largely from precipitation. A considerable amount of water, however, is constantly seeping (at a slow rate) into most rivers from the ground. Groundwater seepage is important in most rivers to sustain perennial flow.

Living Organisms: The total amount of water stored in living organisms is extremely small when compared with other parts of the hydrologic system, but the volume of water circulating through living organisms and back into other parts of the hydrologic system is a different matter. Calculations indicate that in any given period of time, plants may well release as much water into the atmosphere as is discharged by all of the world's rivers combined. This is a significant amount. Its importance is further emphasised when we consider that the residence time of water in organisms is very short, ranging from a few hours, in warm-blooded animals, to a season, for most plants. Thus, living organisms are involved with the flow of a significant volume of water over a period of time, and plants and animals constitute an important segment in the hydrologic system. If we continue to destroy the forests, the effect on the hydrologic system may be devastating.

Surface of Earth: Balance of Water

The water is constantly in circulation from the sea and land to atmosphere in different shapes, which after a certain period of time reaches back to the land and the oceans in the form of precipitation. A water balance can be established for any area of earth's surface by calculating the total precipitation input and the total water output. The total precipitation (input) is equal to the total output of water which is known as *water balance* of the earth's surface. The water balance is a portrait of hydrologic cycle at a specific site or area for any period of time, including estimation of stream flow. Charles Warren Thornthwaite (1899-1963), a pioneer geographer in applied water resource analysis, worked with others to develop a water balance methodology. They related water balance concepts to geographical problems, especially to accurately determining irrigation quantity and timing. The water balance or the water budget may be calculated with the help of the following equation:

$$P = ET + dST + S$$

where, P is the income from precipitation, ET is the loss by evapotranspiration, dST is the gain or loss of storage in the soil, and S is surplus. Thus, all of the income is accounted for by expenditures, an increase or decrease in saving, and a possible surplus that requires wise management. The transfer of moisture to air by evapotranspiration is a function of both available water and available energy. The amount that could be evaporated and transpired by the available energy is the potential evapotranspiration (PE), or simply the water need.

The amount lost upward from the surface is the actual evapotranspiration (AE). It is equivalent to ET in the water budget equation and therefore is limited by precipitation and soil moisture storage. When total precipitation exceeds potential evapotranspiration, part of the water may be needed to restore soil moisture; the remainder, if any, is surplus that may run-off, percolate to the groundwater level, or accumulate in ponds. If PE exceeds AE and soil moisture is being depleted, there is a deficit (D), and the equation becomes as follows:

$$P = PE - dST - D$$

Expenditures fail to meet the need imposed by the energy-moisture system. Obviously, the income is inadequate, although savings may help to offset a temporary shortage. Table below shows an example of water budget accounting on a monthly basis for a representative station in a humid subtropical climate:

Monthly Water Budget: Memphis, Tennessee (all values in millimetres)

	J	*F*	*M*	*A*	*M*	*J*	*J*	*A*	*S*	*O*	*N*	*D*
P	134	112	131	126	106	94	84	80	72	72	107	114
PE	5	10	31	65	113	152	173	160	114	64	26	10
ST*	300	300	300	300	293	241	179	137	119	127	208	300
dST	0	0	0	0	-7	-52	-62	-42	-18	+ 8	+ 81	+92
AE	5	10	31	65	113	146	146	122	90	64	26	10
D	0	0	0	0	0	6	27	38	24	0	0	0
S	129	102	100	61	0	0	0	0	0	0	0	12

Amount of soil moisture storage at end of each month, assuming a capacity of 300 mm.

The water budget concept can be applied in the analysis of the moisture exchange at any surface — a plant leaf, a forest canopy, a glacier, an entire continent, or an ocean. It is an effective means of determining drought conditions and the proper use of irrigation water One of the most fruitful applications is in the predictions of run-off (surplus) from river basins under varying climatic conditions.

Similarly, ground water supplies (another form of surplus) may be calculated. Any single element in the budget can be determined mathematically if all the other elements are known.

Robert W. Christopherson also developed a water balance equation in his monumental work *Elemental Geosystem* in the following way:

PRECIP = POTET - DEFIC + SURPL + STRGE

The terms used in the equation need a definition and explanation.

Precipitation: The particles of liquid water or ice that fall from the atmosphere and may reach the ground is precipitation (PRECIP). It includes rain, sleet, soft hail, hail, mist, fog, dew, etc. About 22 per cent of the earth's precipitation falls on land. Precipitation is measured with the help of rain-gauge, which receives precipitation through an open top so that water can be measured by depth, weight or volume. According to the World Meteorological Organisation, more than 40,000 weather monitoring stations are operating worldwide.

Potential Evapotranspiration: Evaporation is the process in which water in liquid or solid state passes into the vapour state, while the combined water loss to the atmosphere by evaporation from the soil and transpiration from plants is known as *evapotranspiration.* Both evaporation and evapotranspiration are water balance outputs and directly respond to temperature and humidity. Transpired quantities can be significant: on a hot day, a single tree may transpire hundreds of litres of water. Evaporation and transpiration are combined into one term, i.e., evapotranspiration (14% of evaporation and transpiration occurs from land and plants).

Potential Evapotranspiration (POTET): Potential evapotranspiration (water need) is an ideal or hypothetical rate of evapotranspiration estimated to occur from a complete canopy of green foliage of growing plants continuously supplied with all the soil water they can use, a real condition reached in those situations where precipitation is sufficiently great or irrigation water is supplied in sufficient amounts. *Lysimeter* is an elaborate device for measuring potential evapotranspiration.

Deficit (DEFIC): It is also known as *soil water shortage.* It is the soil water budget, the difference between water use and water need, the quantity of irrigation water required to achieve maximum growth of agricultural crops. By substracting DEFIC from POTET, we determine the actual evapotranspiration (ACTET), that takes place. Under ideal conditions POTET and ACTET remain almost same, so that plants do not experience a water shortage, whereas prolonged deficit may lead to drought.

Surplus (SURPL): If potential evapotranspiration (POTET) is satisfied and soil moisture is full, then additional water input becomes surplus (SURPL), or water oversupply. This excess water may sit on the surface, or flow through the soil to groundwater storage. Surplus water that flows across the surface towards stream channels is termed as 'overland flow'. Such flow combines with other precipitation and sub-surface flows into river channels to make up the total run-off from the area — in essence the surface water resource. Because stream flow, or run-off, is generated mostly from surplus water, the water balance approach is a useful tool for indirectly estimating stream flow.

Change in Soil Moisture Storage (STRGE): Soil moisture storage (STRGE) is a 'savings account' of water that can receive deposits and allow withdrawals as conditions change in the water balance. STRGE refers to the amount of water that is stored in the soil and is accessible to plant roots. Rainwater, snowmelt, and irrigation water provide soil moisture recharge.

The texture and structure of the soil dictates the rate of soil moisture recharge. Remember, soil moisture storage is a water savings account with withdrawals (utilisation) and deposits (recharge).

Groundwater Resources

The sub-surface water occupying the saturated zone and moving under the force of gravity is known as *groundwater.* As stated above, groundwater is also a part of the hydrological cycle that lies beneath the surface but is tied to surface supplies. Groundwater is the largest potential source of fresh water in the hydrologic cycle – larger than the surface lakes and streams combined. Between earth's land surface and a depth of 4 km (13,100 feet) worldwide, some 22 per cent of earth's fresh water is found. Despite this volume and its obvious importance, groundwater is widely abused. In many areas it is polluted or consumed in quantities beyond natural recharge rates. In the states of Maharashtra, Rajasthan, Punjab, Haryana and western Uttar Pradesh, the farmers are overpumping the underground water as a result of which the groundwater table is being lowered steadily.

As water is pumped from a well, the adjacent water table within an unconfined aquifer will draw down, or become lower, if the rate of pumping exceeds the horizontal flow of water in the aquifer. The resultant shape of the water table around the well is called a *cone of depression.*

Aquifers are frequently pumped beyond their flow and recharge capacities, resulting in *groundwater mining.* A serious adverse effect of overpumping of groundwater is that the rock layer loses its internal support – the water in the pore spaces between the rock grains – and overlying rock may crush the aquifer, resulting in land subsidence. When aquifers are overpumped near the ocean, the groundwater contact between fresh water and sea water often migrates inland. As a result, wells may become contaminated with salt water, and the aquifer may become useless as a fresh water source.

Significance of Soil Moisture

The role of soil moisture in the water budget is highly variable. The soil moisture depends not only on the other budget factors but also on the capacity of a soil to hold water gained by infiltration. The amount of water a saturated soil can retain against the pull of gravity is its field capacity. It varies mainly with soil texture and also with structure, organic matter content, and the depth of the soil. For example, the fine clays have high field capacities, whereas sandy soils hold little moisture. For agricultural purposes, it is useful to consider the minimum amount of soil that is necessary in the root zone to allow extraction by plants. The minimum is the wilting point, and it depends on the same factors that govern field capacity.

The water content of soil can be ascertained by several methods. A simple, but time-consuming, procedure is to weigh soil samples before and after drying in oven. Unfortunately, the samples cannot be reused. Weighing *lysimeter* accurately measures changes of soil moisture in a specific volume of soil. Another device, the soil moisture *tensiometer,* is a porous clay cell filled with water and placed in the soil.

A pressure gauge attached to the cell registers changing tension as water moves through the cell towards drier soil or into the cell from wet soil. An electrical method depends on the relationship between soil moisture and soil conduction of electricity by buried resistance blocks made by gypsum, nylon or fibre glass. It is specifically useful for remote recording.

When the soil zone has reached field capacity, excess soil water percolates downward under the force of gravity through a sub-surface zone of aeration (vadose water) to a zone of saturation (groundwater). Depths to the upper surface of groundwater, that is, the water table, can be measured in wells and are used to estimate groundwater supplies. Groundwater may also be treated as part of the surplus in the water budget.

It may be observed from table that the highest average precipitation, i.e., 163 cm per year, is recorded in the continent of South America, followed by Africa (69 cm/year) and North America (66 cm/year). Europe and Asia receive 64 cm and 60 cm precipitation per year, respectively, while Australia has the lowest average annual precipitation, being only 47 cm.

The evaporation rate is also highest in South America (70 cm/annum), followed by Africa (43 cm/annum) and Australia (42 cm/annum). The annual evaporation rates in Europe and North America are 39 cm and 32 cm, respectively. Asia, with 31 cm per year, has the lowest rate of evaporation.

Estimated Average Water Budgets of Major Continents and Oceans (cm per year)

	Precipitation	Evaporation	Run-off	Interocean flow
Continent				
Africa	69	43	26	
Asia	60	31	29	
Australia	47	42	5	
Europe	64	39	25	
North America	66	32	34	
South America	163	70	93	
Total land	73	42	31	
Ocean				
Atlantic	89	124	23	-12
Indian	117	132	8	-7
Pacific	133	132	7	+8
World oceans	114	126	12	0

In general, the average annual precipitation and evaporation rates on land (continents) are 73 cm and 42 cm, respectively and the average run-off is 31 cm per annum.

So far as the oceans are concerned, the highest precipitation (133 cm/year) is recorded in the Pacific Ocean, followed by the Indian Ocean (117 cm/per year). The Atlantic Ocean receives only 89 cm of precipitation annually. The evaporation rates in the Pacific Ocean and Indian Ocean are 132 cm/per year each, followed by the Atlantic Ocean, which has the evaporation rate of 124 cm per year. On an average, 114 cm precipitation is received in world oceans and 126 cm of evaporation takes place from them annually.

Flow at Fast Pace

'Ocean current' is a term referring to the faster-moving flows of water in the ocean, as distinct from the slower-moving flows, which are termed as *drifts*, e.g., the *'west wind drift'*. In other words, ocean current is the general movement of a mass of oceanic water in a definite direction which is more or less similar to water streams (rivers) flowing on the land surface of the earth. The land surface heats more rapidly and more intensely than the surface of ocean water body, yet there are meso-level variations in the density and temperature of oceans. These variations lead to the formation of ocean currents. At a global scale, ocean currents help in maintaining the earth's heat-balance by transferring heat from lower to higher latitudes. According to one estimate, ocean currents account for about a quarter of the total heat transport in the oceans, and winds the remaining three-quarters.

Currents of Indian Ocean

The ocean currents of the Indian Ocean may be examined under: i) ocean currents of the southern Indian Ocean, and ii) the ocean currents of the northern Indian Ocean.

The ocean currents of the southern Indian Ocean are relatively simple, a west-east circumpolar drift, mainly under the influence of northwest (anti-trade) winds. Thus, in the southern Indian

Ocean, the circulation broadly resembles that of the other southern oceans, in that it is anti-clockwise.

The ocean currents of the northern Indian Ocean are, however, very closely influenced by the monsoon winds. There is a complete *reversal* in the direction of ocean currents as a result of monsoonal change of air streams.

South Equatorial Current: The Indian south equatorial current has its origin around 10°S. It is probably strengthened by water from the corresponding Pacific current, which has made its way through the Indonesian archipelago. It flows westward to the African coast and turns southward along both the coasts of Malagasy (Madagascar). Being obstructed by Madagascar it is divided into two branches. The northern branch, which enters the Mozambique Channel, is known as the *Mozambique current,* while the one which moves southward along the eastern coast of Madagascar, is known as the *Agulhas current*. The surface temperature of the current remains around 25°C and its average speed is about 20 km per day.

Mozambique Current: It is the northern branch of the south equational current which enters the Mozambique Channel around 10°S. Moving southward between Mozambique and Madagascar, it joins the Agulhas current (warm) around 30° south. Merging with Agulhas, it continues southward along the coast of South Africa upto 35°S, where from it is diverted eastward under the influence of the westerlies (anti-trade winds). Its surface water temperature remains around 22°C and its speed about 20 km per day.

Agulhas Current: It is a branch of the south equatorial current which flows along the eastern coast of Madagascar. It continues southward upto about 30°S where it merges with the Mozambique current. Reaching 35°S, it adopts an easterly movement under the influence of the westerlies (anti-trade winds). Its surface water temperature is generally 25°C and the speed about 25 km per day.

Southwest Monsoon Current: In the summer season, in the northern parts of Indian Ocean, the monsoon winds blow from southwest to northeast. Consequently, the ocean currents follow the general direction of summer monsoon.

Northeast Monsoon Current: In the winter season, the monsoon winds blow from northeast to southwest. Influenced by the direction of winds the ocean currents flow from northeast to southwest. In brief, in the Indian Ocean, the ocean currents are closely controlled by the southwest and northeast monsoons.

Indian Counter-current: In the winter season, under the impact of northeast monsoon, a counter-current develops around 5°S. It flows from Zanzibar to Sumatra.

West Wind Drift (Cold): In the southern parts of the Indian Ocean, around 40°S is a cold water movement from west to east. It develops under the influence of the westerlies (roaring forties) and is known as the *west wind drift.* Reaching 110°E it bifurcates into two branches. One of the two branches turns northward and flows along the western coast of Australia. This is known as the *west Australian cold water current.* Moving northward, it merges into the equatorial current. The second branch of the west wind drift flows to the south of Australia and finally merges into the Pacific west wind drift.

Currents of the Arctic Ocean: The Arctic Ocean is virtually stagnant, almost enclosed basin of water. However, there does seem to be a slow surface drift across the pole from the Siberian coast to that of eastern Greenland. Some cold surface water creeps through the channels between the Canadian Islands into Baffin Bay, contributing to the Labrador current, and other water flows southwards as the East Greenland current.

Currents of the Antarctic Ocean: The movement of water is the Antarctic Ocean is in one sense relatively simple, i.e., a west-east circumpolar drift, mainly under the influence of the northwest winds, with offshoots flowing northward into the other oceans. The surface water, however, moves towards equator, and water from equator moves towards the pole. Thus, the surface and the sub-surface currents have a very complicated relationship.

Small Currents

The water currents of the Mediterranean Sea, Red Sea, and Black Sea are generally known as the *minor currents.* In the Mediterranean Sea a surface current flows eastward through the

Strait of Gibraltar. Similarly, in the Red Sea, a surface current flows from the Arabian Sea through the Bab-el-Mandeb. In the Black Sea, a saline current creeps from the Mediterranean Sea along the bottom of the Bosporus.

Oceanic Currents: Social Effects

Ocean currents closely influence the distribution of temperature, pressure, winds and precipitation, which directly or indirectly influence the economy and society of the people, especially those living in the maritime and coastal regions. Some of the important effects of ocean currents have been mentioned as below:

1. Ocean currents are an important source of temperature distribution. Surface currents distribute tropical heat worldwide. Warm water flows to higher latitudes, transfers heat to the air, cools, moves back to low latitudes, absorbs heat again, and the cycle repeats. The greatest amount of heat transfer occurs at mid-latitudes, where about 10^{15} (10 million billion) calories of heat are transferred each second – more than a million times the power consumed by all the world's human population.
2. The ocean currents mitigate the air temperatures and help in the regulation of air routes. Commercial airlines flying to Europe from USA west coast cities, fly over central Ontario. In winter and spring months, the ground in USA remains hidden beneath masses of ice and snow, but passengers can often make out the frozen surface of James Bay. When these airlines land in London, their passengers find a much milder climate than the barren whiteness of central Canada. Yet London's latitude of 51°N is the same as the southern tip of James Bay. This difference in the weather and climate of the given two places (James Bay and London) is because the wind that blows from America towards Europe becomes relatively warm after coming into contact with the warm water of Gulf Stream and North Atlantic Drift. Ireland, Wales, Scotland, England and the West European countries, therefore, have a maritime climate. The cities like London and Edinburgh are warmed by the energy of tropical sunlight transported to their high latitudes by the Gulf Stream.

3. At lower latitudes, on the eastern boundaries of oceans, the situation is often reversed. The eastern parts of ocean in lower latitudes record relatively low temperatures making the weather and climate cold, and foggy. In the summer season (July and August), San Francisco (west coast city of USA) has cool, foggy and mild weather, while Washington, D.C., on nearly the same latitude (but on the eastern boundary of an ocean basin) is known for its August heat and humidity. This striking variation in the weather of these two places situated almost at the same latitude is simply because of the role of ocean currents in the transmission of heat from equator to poleward and the cold water from polar regions towards the lower latitudes.
4. The convergence of warm and cold waters near Newfoundland and near Japan results into foggy conditions which are hazardous to navigation. The same conditions are, however, most suit able for the fast growth of plankton (a food for fish). Consequently, the Grand Bank of North America and Easter coast of Japan are the leading fishing grounds of the world.

Significance of Tides

The regular rise and fall of water level in the world's seas and oceans, resulting from the gravitational attraction that is exerted upon the earth by the sun and the moon is known as *tides*. The tidal-producing force of the moon is slightly more than twice to that of the sun. In other words, tides are produced by the gravitational attraction of the moon and the centrifugal force of the earth and the moon.

The tides are of different magnitude and their height in the open sea differs from that in the coastal areas. In the open ocean, the difference in height between high (spring) and low (neap) tides, i.e., the range, may be only half a metre, but in shallow marginal sea it can increase to about 10 metres (33 feet) and in some of the estuaries up to 12 metres (40 feet). An average tide rises about 4 metres.

Spring Tide: A tide with a range considerably increased from that of the mean tidal levels. It occurs twice each month around the time of the new moon *(conjunction)* and the full moon *(opposition)*. It is due to the complementary gravitational effects caused with the earth.

Neap Tide: A tide occurring near the lunar quardrature during which the tidal-producing forces do not supplement each other, thereby causing relatively small tidal ranges. Thus, high tides are somewhat higher (by some 10-30%) than average, all of which slow down the velocity of tidal currents at this time (about every 14.75 days).

Further variations are produced by the position of the moon relative to the earth. When it is at the nearest point (or *perigee)* its tide-producing effect is more pronounced, and these perigean tides are about 20 per cent higher at high tide. If they coincide with spring tides, very large tidal ranges occur. Conversely, apogean tides, when the moon is at its farthest distance, are lower than usual at high tide, and if these coincide with neap tides, the range is small. The highest spring tides of all occur at the equinoxes, known as the *equinoctial springs.*

Tide Producing Forces

The moon, the sun and the centrifugal force of the earth are the main tide-producing forces. In general, a heavenly body exerts a gravitational effect on the earth's surface, the strength of which varies directly with its mass and inversely with the square of its distance. Thus, the sun's mass is 26 million times that of moon, but it is 380 times farther away, and its tide-producing effect on the earth is only 0.46 times that of the moon. When the moon, the sun and the earth are in the same line, either in conjunction or in opposition (positions known as in *syzygy),* their attraction is complementary and so the highest high tides and lowest tides occur.

Effect of Tides

Tides have great importance for man and society. Tides increase the depth of water in shallow harbours and over the shallow entrances of some harbours, thus making the port available during high tide. Moreover, they scour out the entrances of inlets, remove wastes dumped into the sea from cities, and distribute the mud and silt brought to the ocean by rivers. They modify the contour of shorelines, and affect the human use of coast in many other ways. Tides are also being used in the developed countries like USA, France, UK, Germany, Italy, Japan, etc., for the generation of hydro electricity. The first major tidal power station was opened in France in 1966.

Tidal Origin: Different Concepts

A number of theories have been put forward about the origin of tides. The important theories about their origin are: i) the Equilibrium Theory, ii) the Progressive Wave Theory, and iii) the Stationary Wave Theory.

The Equilibrium Theory: This theory was put forward on the basis of Newton's law of gravitation (1686), which states that each body in the universe attracts every other body with a force directly proportional to the product of their masses and inversely proportional to the square of the distance between them measured from their centres of mass along a line joining these centres. In other words, the celestial bodies attract each other through their gravitational force in such a way that they remain in equilibrium. The sun, the moon, and the earth are also in equilibrium due to their respective pull towards each other. Though the gravitation force of the sun is far greater than that of the moon, but the lunar gravitational force has more effect on the earth than the sun because of its nearness to the earth. As a result of this, the water of the earth's surface under the moon is attracted and pulled and high tide is caused. The opposite side of the earth's surface also experiences tide because of the centrifugal or reactionary force. Because of the force of gravitation, the highest point of rise of water (high tides) lies nearest to and farthest away from the moon, while the lowest points of the water surface (lowest tides) lie at places perpendicular to the above places.

The theory has not been universally accepted as it does not explain some of the points about the occurrence of tides. For example, the earth's surface is composed of land (29%) and oceans (71%). Owing to the nature of composition of the earth's surface, the moon will not be so effective as it would have been, if the earth's surface would have been composed of only water.

The theory also does not explain how the tides occur in the areas of oceans where the horizontal movement of water is absent or insignificant. According to scientists, the bulge of water may not be possible unless some sort of horizontal movement of tide is involved.

In addition to these, one more criticism about the equilibrium theory is about the time of occurrence of tides. For example, the time of high tide should be the same at all places along each meridian but this never happens.

In view of the above weaknesses Airy opposed this theory and declared it erroneous as it does not explain the origin of tides as a result of the gravitational force.

Theory of Progressive Waves: This theory was propounded by William Whewell in 1883. It is based on the following facts:

- The earth is a heterogeneous body and not a perfect fluid (surrounded by water on all sides).
- Tides occur at different times at different places on the same longitude.
- There is a lagging of time of tides away from the source.
- There is a variation in the intensity of tides at different places.
- Tide is in the form of tidal wave which travels from east to west. The crests and troughs of such tidal waves become tides and ebbs, respectively.
- The tidal waves are originated in the oceans under the influence of tidal force of the moon.
- The length and speed of the tidal waves depend on the depth of seas and oceans.
- In a globe completely surrounded by water, the tidal waves would travel freely from east to west but the position of land and water hinders the speed and direction of these waves.
- Since the continents roughly stretch from north to south, they hamper the free movement of tidal waves. These waves

are least hampered in the oceans surrounding the Antarctic Ocean, owing to the non-availability of land in the higher latitudes of the Southern Hemisphere. Because of the above facts, the tidal waves are generated in the oceans of the Southern Hemisphere, under the influence of tide-producing force of the moon. These waves are called as the *primary waves* which move from east to west in the form of forced waves.

These primary waves are obstructed by the continents and are consequently refracted northward. Consequently, the *secondary waves* are generated when the westward movement of primary waves is obstructed by land masses. The northward moving secondary waves are also called as the *derived waves.* Further minor waves are generated from these secondary waves, which may be termed as the *tertiary waves.* The secondary and tertiary waves move northward with decreasing intensity and magnitude but generate tides everywhere. It may be mentioned that the primary waves are influenced by the moon but the tertiary or minor waves move freely.

Thus, the tidal waves after being originated in the oceans of Southern Hemisphere, progressively move northward with continuous lag of time and dissipation of wave energy. In other words, the arrival of these progressive waves at successive places northward along the same longitude is also progressively delayed. This explains the delay in the occurrence of tide at different places on the same longitude. Thus, the time of tide is progressively delayed northward along the longitude. These progressive waves become insignificant and ineffective after reaching the North Pole. Moreover, the crests and troughs of these waves after reaching the coasts cause tides and ebbs, respectively.

The progressive wave theory is, however, not free from criticism. According to this theory, the age of tides increases northward. In other words, if the tide is generated in the south on a particular longitude, it reaches quite late at the points located further north on the same longitude. Normally, the tides are local or regional phenomena rather than phenomena originating in the southern ocean and moving progressively northward. At some of the latitudes, both the daily and semi-diurnal types of tides are observed. Moreover, there is spatial variation in the irregularity of tides in different oceans. These variations cannot be explained on the basis of the progressive wave theory.

Concept of Stationary Wave

This theory was propounded by R.A. Harris of the Geodetic Survey of USA. This theory, which was developed as a reaction to the progressive wave theory, gives a satisfactory explanation for the locational differences and variations in tides.

In the opinion of Harris, the phenomenon of tide is not due to progressive waves which originate in the oceans of Southern Hemisphere as claimed by Whewell but because of the stationary waves which originate independently in each ocean. In other words, the tides are regional phenomena.

In the huge water bodies of oceans, the sun and the moon cause oscillations, but the oscillation does not occur along straight lines. This process results in the formation of waves. Every stationary wave has a definite time of its oscillation. The oscillation system and process are affected by the depth, configuration, length and breadth of the ocean basin. These waves, after their origin, move towards the coasts. The forward movement of these waves is, however, hampered by the continental peninsulas, islands bays, gulfs and straits. Reaching the coasts, the crests and troughs of the waves cause tides and ebbs, respectively. Thus, there is a positive correlation between the depth of the oceans and the height of the tides. In other words, greater the depth of the ocean, higher the stationary waves generated which lead to high or spring tides. Low tides are caused in the shallow seas, because of the lower heights of the stationary waves.

The main advantage of the stationary wave theory lies in the fact that it helps in making reliable prediction about the magnitude of the tides.

Significance of Waves

In hydrology and oceanography, wave means a deformation of water surface in the form of an oscillatory movement which manifests itself by an alternating rise and fall of that surface. The oscillation is generated by wind's pressure on the water surface.

Some waves are, however, caused by sub-marine earthquakes *(tsunami)* or landslides into water bodies.

Ocean waves have distinct parts. The *wave crest* is the highest part of the wave, above average water level, the *wave trough* the valley between wave crests, below average water level. *Wave height* is the vertical distance between a wave crest and the adjacent trough, while *wave length* is the horizontal distance between two successive crests (or troughs).

The height, length and period that are eventually achieved by a wave depend upon three factors: i) the wind speed, ii) the length of time the wind was blown, and iii) the distance that the wind has travelled across the open water. As the quantity of energy transferred from the wind to the water increases, the heights of waves transmitting it increase. In the open ocean, wave heights of 1.5-4.5 metres (5-15 feet) are common although storms may produce much higher waves.

It is important to realise that in the open sea the motion of the wave is different from the motion of the water particles within it. It is the wave form that moves forward, not the water itself. Each water particle moves in a circular path during the passage of a wave. As a wave passes, a water particle returns almost to its original position. The turbulent water created by breaking wave is called *surf*.

In brief, waves transmit energy, not water mass, across the ocean's surface. The speed of ocean waves usually depends on their wavelength, with long wave moving fastest. Arranged from short to long wavelengths (and therefore from slowest to fastest), ocean waves are generated by very small disturbances (capillary waves), wind (wind waves), rocking of water in enclosed spaces (seiches), seismic and volcanic activity or other sudden displacements (tsunami), and gravitational attraction (tides). The behaviour of waves depends largely on the relation between a wave's size and the depth of water through which it is moving. Waves can refract, and reflect, break, and interfere with one another.

Unlike wind waves, the waves of very long wavelength are always in 'shallow water' (water less than half of their wavelength deep). These long waves travel at high speeds. Some of the waves can be destructive, but their ability to cause damage is fortunately not proportional to their wavelength. The waves with the longest wavelength are the tides.

Different Causes

The pattern of oceanic circulation is produced by the interaction of a number of factors. The main factors which produce the ocean currents are:

1. Planetary winds.
2. Variations in sea water temperatures.
3. Variations in sea water salinity.
4. Rotation of the earth.
5. Configuration of coastlines.

The Planetary Winds: The prevailing planetary winds (trade winds, anti-trade winds or westerlies, and the polar winds) play vital roles in the origin and development of ocean currents. Most of the earth's surface energy is concentrated in each hemisphere's trade winds and westerlies. Tiny irregularities in the sea surface, called *capillary waves,* transfer some of the energy from the moving air to the water by friction. This tug of wind on the ocean surface begins a more rapid mass flow of water. As a rule, the friction of wind blowing for at least ten hours will cause surface water to flow downwind at about 2 per cent of wind speed. The water flowing beneath the wind forms a surface current. Because of the Coriolis effect, the Northern Hemisphere currents flow to the right of the wind direction while in the Southern Hemisphere currents flow to the left. Intervening continents and basin topography often block continuous flow and help to deflect the moving water into a circular pattern. This flow around the periphery of an ocean basin is called a *gyre* (gyros = circle). Most of the ocean currents of the world follow the direction of prevailing, permanent or planetary winds. For example, the equatorial currents flow westward at a speed of 40 km (25 miles) per hour, under the impact of the trade winds. The Gulf Stream in the Atlantic Ocean and Kuroshio in the Pacific Ocean move in northeastern direction under the influence of the anti-trade (westerlies) winds. Many of the ocean currents are drifts caused by friction between the winds and the surface water. They move more or less in the direction of the wind and vary in position and strength with the seasonal winds, e.g., Indian monsoons.

Variations in Sea Water Temperatures: There are marked variations in the horizontal and vertical distribution of temperatures of the oceans. In general, the temperature decreases from the equator towards the poles and from the surface towards the bottom of the seas and oceans. Thus, in the equatorial region the density of water decreases due to high temperature. There is a negative correlation between the temperature received and the density of water, which means higher the temperature, lower the density of water, and vice versa. As a result of this the less warmer and lighter water from the equatorial region moves towards the colder and denser waters of the polar areas. Contrary to this, there is a movement of ocean water below the water surface in the form of sub-surface current from colder polar areas to warmer equatorial areas. The Gulf Stream and Kuroshio (warm currents) moving from equator towards North Pole and the Labrador and Kurile currents moving from polar areas towards equator are some of such examples.

Variations in Sea Water Salinity: The amount of salts contained in sea water does vary from one part of the ocean to another. Water with a high salinity is denser than that with lower amounts of salts. The high salinity water tends to subside and move below water of low salinity. Ocean currents on the water surface are generated from the areas of less salinity to the areas of high salinity. There is marked variation in the salinity of the Atlantic Ocean and the Mediterranean Sea. Because of this variation, the ocean current flows from the Atlantic Ocean to the Mediterranean Sea. A similar ocean current may be observed between the Indian Ocean and the Red Sea via Bab-el-Mandeb. An identical pattern is found in the Baltic Sea and the North Sea (Atlantic Ocean). The Peru current may also be cited as an example of a current which has its origin because of the variation in the density of water.

Rotation of the Earth: The earth rotates on its axis, from west to east. This rotation is the cause of deflective force known as *Coriolis force*, which deflects the general direction of the winds and that of the ocean currents. For example, the currents flowing from equator towards the north and south poles are deflected to their right in the Northern Hemisphere and towards their left in the Southern Hemisphere, respectively. The counter-equatorial currents are also the result of rotation of the earth.

Configuration of the Coastlines: The shape and configuration of the coastlines also have a close influence on the direction and movement of the ocean currents. For example, the equatorial current after being obstructed by the Brazilian coasts is bifurcated into two branches. The northern branch is known as the *Caribbean current* flowing along the northern coast of South America, while the southern branch moves along the eastern coast of Brazil which is known as the *Brazilian current.* In the Indian Ocean, the monsoon currents closely follow the coastlines.

In addition to the shape of coastline, the configuration of the bottom relief also modifies the direction of movement of the ocean currents. In other words, the submarine ridges usually deflect the direction of the ocean currents. Generally, the ocean currents while crossing over a submarine ridge are deflected to the right in the Northern Hemisphere and to the left in the Southern Hemisphere. For example, the North Atlantic Drift is deflected to the right when it crosses the Wyville-Thomson ridge. Similarly, the north equatorial current is deflected to the right while crossing over the mid-Atlantic ridge.

Currents of Atlantic Ocean

North and South Equatorial Currents: To the north and south of equator there are two westward moving currents, i.e., the north and south equatorial currents. These currents derive their energy mainly from the trade winds that blow from the northeast and the southeast in the Northern Hemisphere and Southern Hemisphere, respectively, towards the equator. Because of the deflection created by the rotation of the earth (Coriolis force), these currents move almost due west. Between the westward-flowing currents there is a weaker, oppositely directed eastward flow, known as the *equatorial counter-current*. The north and south equatorial currents pile water up against the western coast of South America, and because the trade winds are weak along the equator, some of the piled-up water flows back 'downhill' (eastward), creating the counter current.

As the north equatorial current continues west, the presence of South American continent and the Coriolis force (rotation of the earth) turn the water to the north. The water then moves northwest through the Caribbean to the mouth of the Gulf of Mexico at the Straits of Yucatan. From here some curves towards the right, flowing some distance off the shore of the Gulf of Mexico, and part of it curves more sharply towards the east and flows directly towards the north coast of Cuba. These two parts are known as the *Antilles current* and the *Caribbean current*, respectively. The Antilles current is diverted northward and flows to the east of West Indies islands which helps in the formation of Sargasso Sea eddy, while the second branch (Cribbean Current) enters the Gulf of Mexico to form the Gulf Stream.

The south equatorial current has its origin from the western coast of Africa, from where it moves towards South America. The south equatorial current is more strong, steady and covers a wider extent as compared to the north equatorial current. This warm current is bifurcated into two branches due to the obstruction of land barrier in the form of east coast of Brazil. The northward branch merges with the north equatorial current near Trinidad, while the second branch continues along the eastern coast of Brazil where it is known as the *Brazilian current* (warm water).

Gulf Stream

The Gulf Stream is the largest of the western boundary currents of the North Atlantic Ocean. This is a warm water current which originates in the Gulf of Mexico around 20°N and moves in a northeasterly direction along the eastern coast of North America. Studies of the Gulf Stream have revealed that near the Miami coast, the Gulf Stream moves at an average speed of 2 metres per second (5 miles per hour) to a depth of 450 metres (1,500 feet). Water in the Gulf Stream can move more than 160 km (100 miles) in a day. The average speed of the current is, however, 33 km (20 miles) per day. Its average width is about 70 km (43 miles). In the Gulf Stream, the current as river analogy can be visible. Water within the current is usually warm, blue, often depleted of nutrients and incapable of supporting much life. By contrast, the continental slope adjacent to the current is often cold, green, and teeming with life. The current meanders as it flows poleward. The looping meander sometimes connects to form turbulent rings, or *eddies,* that trap cold or warm water in its centre and then gets separated from the main flow. Under the impact of westerlies, this warm water current reaches the western coasts of Europe upto 70°N.

Along the coast of Florida the average surface temperature of the Gulf Stream reads about 25°C. Moving northward at 30°N, the surface temperature decreases to about 10°C. The general direction of flow of the Gulf Stream, north of the 30°N latitude, is northward, but beyond *Cape Hatteras* it bends slowly to the right, passing 350 km (210 miles) south of Natucket until south of Halifax the flow is nearly due east. Where the Gulf Stream leaves the continental shelf at Cape Hatteras, its average width is about 82 km (50 miles). Eastward, it widens gradually, becoming 120 km (about 70 miles) wide in the longitude of Halifax.

The width of the Gulf Stream varies in different latitudes. The temperature of water near the coast upto 40°N ranges between 4° and 10°C. This zone of cold water between the coast and Gulf Stream is known as *Cold Wall.* Moving along the east coast of the United States, it is strengthened by the prevailing westerly winds and is deflected to the east between 35°N and 45°N latitudes. Near Newfoundland its water mixes with that of the cold water current of Labrador which results in the formation of dense fog. The dense foggy conditions around the Newfoundland are hazardous to the navigation of ships.

As the Gulf Stream continues north-eastward beyond the Grand Bank, it gradually widens and decreases in speed until it becomes a vast, slow-moving current known as the *North Atlantic Drift.* As the North Atlantic Drift approaches western Europe, it splits. Part of it moves northward, past Great Britain and Norway. The other part is deflected southward as the cool Canaries current.

The warm water current of Gulf Stream modifies the weather conditions of the eastern coast of USA and Canada, and the western coast of Europe. The temperature of the eastern coast of USA becomes significantly high during the summer months. On the western coast of Europe, the warm water of this current keeps the sea ports open even in the severe winter season and make the climate and weather milder.

Different Warm and Cold Currents

Labrador Current (Cold): The Labrador, an important cold water current of the North Atlantic Ocean, has its origin in the Arctic Ocean; it flows from north to south between the Greenland and the Baffin islands (Davis Strait). Passing southward, it merges with the Gulf Stream near Newfoundland. The average speed of the current is about 28 km (15 miles) per day. It brings down huge icebergs from the Arctic Ocean to the eastern coast of Canada which are hazardous to navigation. The convergence of this cold water current with the Gulf Stream near Newfoundland results into dense fog. The merger of these currents is, however, considered as most conducive for the fast growth of planktons which is a feed for fish. It is because of these favourable conditions that the Grand Bank near Newfoundland has become as an ideal fishing ground of the world.

Canary Current (Cold): The Canary is a cold water current which flows along the western coast of North Africa. The average speed of the current is about 2 km (1.2 miles) per hour. The current is so shallow and broad that sailors may not notice it. It brings cold water of the higher latitudes towards the equator upto 15°N. The surface temperature of the current varies between 10°C along the Spanish coast to 20°C to the west of Canary Islands at 20°N. The relative coolness of the Canary current ameliorates the temperatures and weather in the lower latitudes along the west coast of Africa. It, however, reduces the relative humidity of the air mass and winds that move eastward from the Atlantic Ocean. This is one of the main causes of scanty rainfall in the greater parts of north-central Africa. Thus Canary current helps in the desertification of Sahara Desert.

Sargasso Sea: The word Sargasso Sea has been derived from the Portuguese word 'sargassum' meaning sea weeds. The Sargasso Sea has abundance of weeds and the highest salinity of the Atlantic Ocean, being 37 per cent. The areal extent of Sargasso Sea is between 20°N to 40°N and 35°W to 75°W. The mean annual temperature of the sea is 28°C. The sea is infested with floating weeds which obstruct navigation.

Brazil Current (Warm): Along the eastern coast of Brazil flowing from north to south upto 40°S is the warm water current of Brazil. The surface temperature of the current reads about 25°C in the north and about 15°C at 40°S. The average speed of the current varies between 28 km (15 miles) per day. It ameliorates the weather conditions along the eastern coast of Argentina.

Falkland Current (Cold): The Falkland current has its origin in the Antarctic Ocean around 65°S. It flows from south to north. The average speed of this current is about 18 km (10 miles) per day. It brings large-sized icebergs from the Antarctic Ocean to make the Falkland and eastern coast of south Argentina cold.

Benguela Current (Cold): The Benguela is a cold water current which has, its origin in the Antarctic Ocean and flows along the coast of southwest Africa, generally between Cape Town and 18°S. It is characterised by the upwelling of relatively cold water. The average temperature of the surface water of the current varies between 13°C and 17°C. The average speed of this current is about 18 km (10 miles) per day. The upwelling water of this current cools the western coast of South Africa and Namibia. The desert of Kalahari is largely the result of this current as it helps in reducing the relative humidity of the eastward moving warm and moist air masses.

South Atlantic Drift (Cold): The eastward movement of the Brazilian current is known as the *South Atlantic Drift*. It develops at 40°S owing to the impact of the westerlies which are known as the *roaring forties*. Consequently, it is also known as the *westerlies drift* or the *Antarctic drift*.

Currents of the Pacific Ocean: It may be observed that the same broad circulatory systems, clockwise in the Northern Hemisphere and anti-clockwise in the Southern Hemisphere, are present in the Pacific Ocean also. Some of the major ocean currents of the Pacific Ocean have been briefly described as under:

North and South Equatorial Currents: The north equatorial current flows westward which originates around 10°N to the west of Mexico (North America). Moving westward, almost parallel to the line of equator, it reaches the coasts of Philippines after covering a distance of about 12,000 km (7,500 miles). The Pacific Ocean is too wide in the lower latitudes and, therefore, a greater volume of water is involved. After reaching the coast of the Philippines, under the impact of Coriolis force, it takes a northerly direction. Its surface temperature varies between 25°C and 30°C. The speed of the current ranges between 10 km and 25 km per day.

Under the impact of southeast trade winds, the south equatorial current originates to the west of Peru around 10°S latitude. It is one of the strongest currents. The speed of the current varies between 16 km (10 miles) to 30 km (18 miles) per day. The surface temperature of its water reads between 20°C and 25°C.

Both the north and south equatorial currents flow westward, with a compensatory counter-current flowing in the reverse direction between them along a line about 5°N. The counter-equatorial current reaches upto Panama.

Kuroshio Current: Kuroshio is the most important warm water current of the North Pacific Ocean. It is analogous to the Gulf Stream of the Atlantic Ocean. It develops partly due to the Coriolis force and partly due to the obstruction of the Philippines in the flow of the north equatorial current. The average surface temperature of this current remains around 18°C. It moves at a speed of about 30 km (18 miles) per day. It keeps the eastern coast of Japan warm even in the coldest month (January) when snowing is frequent in Honshu and Hokkaido.

An offshoot of Kuroshio current, also known as *Tsushima current*, enters into the Sea of Japan along the west coast of the islands. The relatively warm water of Tsushima current keeps the western coast of Japan warm. Around 35°N, the Kuroshio current, under the impact of westerlies, leaves the coast of Japan and adopts a northeasterly direction, and reaches the western coast of North America around 150°W. Further north it is known as the *Aleutian current*.

Kurile or Oyashio Current: The Kurile current originates from the Bering Strait and moving southward along the coast of Kamchatka it touches the island of Kurile, where from it is called as the *Kurile current*. It carries cold water and icebergs from the

Arctic ocean to the coast of Kurile (Russia) and Hokkaido (Japan). Between Sakhalin and Hokkaido, this cold water current is known as the *Oyashio current*. Passing Hokkaido, it sinks, like the Labrador current, beneath the warm waters of the Kuroshio or the North Pacific drift. To the north and east of Japan, its cold water converges with the warm water of Kuroshio current. The merger of these two currents results into dense fog like that of Newfoundland which is hazardous to navigation. These conditions are, however, ideal for the abundant growth of plankton (a food for fish). Because of these conditions, the northeastern coast of Honshu and Hokkaido is one of the most important fishing grounds of the world.

California Current: It is a cold current which flows southward along the Pacific coastline of USA, caused by the upwelling of colder water from greater depths due to the southward deflection of the North Pacific current and the transference of surface water westward across the Pacific Ocean as the north equatorial current. This cold current causes a marked equatorward bend of the global isotherms and has a marked effect on the coastal climate of Oregon and north California. It results into the heavy sea fogs off San Francisco.

Peru Current (Humboldt Current): The Peru current, also known as the *Humboldt current*, is a cold water current which flows along the western coast of South America. It is caused by the upward welling of colder water in response to the northward deflection of the west wind drift and the transference of surface water westward across the Pacific Ocean as the south equatorial current.

The mean annual surface temperature remains around 15°C and the speed of the current is about 27 km (15 miles) per day. This cold water current causes a marked equatorward bend of the global isotherms and has pronounced effect on the coastal climate of Chile and Peru.

East Australian Current (Warm): It is a branch of the south equatorial current which moves from north to south along the eastern coast of Australia. Around 40°S, under the impact of Coriolis force, it is deflected towards east and touches the coasts of New

Zealand. It raises the temperature along the eastern coast of Australia and the coasts of New Zealand.

West Wind Drift (Cold): To the south of Tasmania and New Zealand flows water almost from west to east, which is known as the *west wind drift.* This water drift is largely confined between 40°S to 50°S latitudes. Under the influence of roaring forties this current achieves a great speed reaching upto 30 km per day.

Wealth of Water

Conditions for Development of Karst

There are four conditions which contribute to the maximum development of karst. These are:

1. There must be present at or near the surface a soluble rock, preferably limestone. The limestone must be massive, thickly bedded, hard, tenacious and well cemented.
2. The soluble rock should be dense, highly jointed, and preferably thinly bedded.
3. The third condition which favours an excellent development of karst is the existence of entrenched valleys below uplands underlain by soluble and well-jointed rocks. This favours ready downward movement of groundwater through the rock.
4. The region must be one of at least moderate rainfall. In general, arid and semi-arid regions do not exhibit marked development of karst. It is significant that nearly all the notable karst regions of the world are in the areas of moderate to abundant rainfall.

The work of underground water is important in the areas of limestone rocks where it gives rise to distinctive landforms. Like running water, underground water also erodes, transports and deposits.

Erosional Landforms: The following are the main erosional landforms of a karst region:

Karren: 'Karren' is a German term initially used to describe minor and major solution furrows or funnels cut by surface water (groundwater) into limestone. Now, the term is used for highly corrugated and rough surface of limestone lithology, characterised by low ridges and pinnacles, narrow clefts and numerous solution holes. In French, it is called as *lopies.*

Sinkhole: A funnel-shaped or cylindrical-shaped hole, depression, or sink in the ground surface of a limestone or chalk terrain. It is equivalent to a *ponor* of karst country.

It is usually dry or exhibits only minor seepage of surface water and should be distinguished from a *swallet,* which marks the disappearance of a surface stream. A sinkhole is formed by subterranean collapse of a cave system or by surface solution. The depth of the sinkhole varies from a few centimetres to about 10 metres. Some swallow holes are further enlarged due to continuous solution into larger depressions which are called *dolines.*

Caverns (Caves): Caverns or caves are also one of the important characteristic features of groundwater in limestone regions. Caverns are formed in several different ways. The rocks in which most caves occur are salt, gypsum, dolomite and limestone, with the latter by far the most important.

Groundwater charged with carbon dioxide and other chemicals seeps along stratification and joint planes or other cracks in the rock, and slowly dissolves the soluble rocks. Wholly stagnant groundwater soon become saturated with minerals and loses its effectiveness in the cave-making process. It is, therefore, necessary that there should be such a condition of underground drainage that the saturated water can move away with its chemical load and be replaced with a new supply of water which is capable of continuing the solvent action. The underground drainage is most effective above the water table, but there is abundant evidence of groundwater movement well below this table.

As the solvent action goes on, the caves continue to grow, usually they are irregular in size and shape. Some are huge rooms, others are vast labyrinths of intricate branching passages. In some caves there are pools of standing water, in others, there are large streams of moving water. The Holloch cave with a length of 85 km is the longest cave in the world, followed by Fint Ridge cave (81 km) and Mammoth cave (72 km) of Kentucky.

Terra Rossa: When rainwater dissolves part of surface rock and enters the sub-surface, particles of red clay soil are deposited on the surface as well as in the opened joints. This is called *terra rossa* which resembles to lateritic soil. It may not be present at steep slopes but can be seen in areas which are either flat or have gentle slope. Sometimes it may be several metres thick and may entirely cover the rocky surface.

Karst Window: Karst window is formed due to collapse of upper surface of sinkholes or *dolines.* These windows enable the researchers and geomorphologists to observe sub-surface drainage and other features formed below the ground surface.

Uvala: A large surface depression (several km in diameter) in limestone terrain (karst region) is known as *uvala.* It is formed by the coalescence of adjoining dolines. It has an irregular floor which is not as smooth as that of polje.

Polje: Polje is a large depression in a karst region with steep sides and a flat floor. If it is drained by surface water sources, it is termed as *open polje,* but if it drains by means of swallow holes, it is a *closed polje.* The depression is thought to have been formed by the coalescence of collapsed cave systems. Its floor is generally covered with alluvium. The Livno polje of the Balkan region of Europe is 64 km long and 5-11 km wide. In Croatia, Serbia and Bosnia-Herzegovina, poljes are devoted to maize cultivation on account of flat surface and easy availability of water.

Swallet or Swallow Hole: The point at which a surface stream disappears underground in a limestone (karst) terrain prior to commencing its underground journey is known as a *swallet.* In France, it is called as *embut.*

Sinking Creek: The surface of the karst plain looks like a sieve because of numerous, closely spaced sinkholes. When surface water disappears through numerous sinkholes located in a line, the resultant feature is called *sinking creek.*

Blind Valley: A type of valley in a karstic limestone terrain. It may be occupied by a stream which disappears underground at the valley's lower end as it approaches an enclosing rock-wall. Consequently, the valley looks a dry valley.

Avert (Ponores): 'Aven' is a French term which has been universally adopted to describe a deep shaft-like hole in limestone terrain, leading down into extensive cave systems (karst).

Natural Bridge: Natural bridge is an erosional feature in karst topography. They are formed either due to the collapse of the roofs of caves or due to the disappearance of surface streams as subterranean streams.

Depositional Landforms: The mineral matter dissolved by groundwater can be deposited in a variety of ways. The most spectacular deposits are stalactites and stalagmites which are found in caves. Less obvious are the deposits in permeable rocks such as sandstone and conglomerates. Here groundwater commonly deposits mineral matter as a cement between grains.

After limestone caverns have formed, it frequently happens that water seeping through the walls or the roofs deposits calcium carbonate in the form of stalactites or stalagmites (sometimes called *dripstone*). The deposition from solution is brought about by: (i) evaporation of the water, (ii) reduction in pressure, and (iii) loss of gases. The main depositional features by these processes are called as *stalactites* and *stalagmites.*

Stalactites: Stalactites are calcite-like forms that hang from the roofs of caves. In other words, a tapering pendant of concretionary material descending from a cave ceiling, created by the reprecipitation of carbonate in calcite form percolating groundwater is known as a *stalactite.*

Stalagmites: A columnar concretion ascending from the floor of a cave. It is formed from the reprecipitation of carbonate in calcite form perpendicularly beneath a constant source of groundwater that drips off the lower tip of a stalactite or percolates through the roof of a cave in a karst environment. It may eventually combine with a stalactite to form a pillar.

An almost endless variety of beautiful and interesting adornments may be formed through deposition of calcite in limestone caves. Many stalactites and stalagmites eventually unite to form columns. Water percolating from a fracture in the roof may form a thin, vertical sheet of rock known as *drip curtain.* Pools of water on the cave floor flow from one place to another, and as they evaporate, calcium carbonate is deposited on the floor, forming *travertine terraces.*

Karst Erosion Cycle: The concept of 'normal cycle of erosion' advocated by Davis was applied in the karst areas by J.W. Beede (1911) and J. Cvijic (1918). According to them, the evolution of landscape in the limestone regions is a special and transitory phase marking the mature stage of the normal cycle in regions where the structure was especially favourable to solution. The basis of such a view arises from the fact that even in limestone regions the cycle starts with surface drainage which also ends with surface drainage. In other words, the karst cycle of erosion may be said to begin with the transfer of surface drainage to underground routes and end with its reappearance on the surface.

There are two ways in which a karst cycle may commence: i) through uplift above base level of a limestone terrain on which fluvial erosion had been in progress, and ii) through uplift of an area of clastic rocks beneath which are limestones above the new base level. In either event, the cycle begins with surface drainage lines but the transformation to underground drainage proceeds by somewhat different methods.

J. Cvijic (1918), on the basis of study of the karst topography of Yugoslavia, recognised the four stages of 'karst cycle': i) youth stage, ii) mature stage, iii) late mature stage, and iv) old stage. These stages are briefly described below:

Youth Stage: The youth stage begins with surface drainage on either original limestone surface or the one that has been exposed on the surface by erosion. Slowly the surface drainage sinks below and there is progressive development of underground drainage. Sinkholes, lapis, and stray dolines develop, but underground drainage is incomplete and there is no formation of larger caverns.

Mature Stage: In the maturity stage, there is a maximum of sinkholes and underground drainage with only major entrenched streams persisting as surface streams, sinking creeks, swallow holes, blind valleys, compound sinks, and dolines by the thousands typify this stage. There is extensive cavern development and cave travertines are formed. In fact, it is the stage when the maximum karst development takes place.

Late Maturity Stage: In the late maturity stage the features of karst topography start declining. These features expand to form uvulas, and gradually poljes diversified by *hums* (isolated hills) dominate the surface.

Old Stage: The old stage is initiated with the beginning of return to surface drainage. The karst windows, karst tunnels, natural bridges, and collapse *hums* (isolated hills) are the most significant features. This is the stage which resembles a peneplain. The second karst cycle of erosion may start with fresh upliftment of karst land.

Significance of Springs

A spring is a natural outflow of water from the surface of the ground. It may flow strongly and even gush out with considerable force, or it may just ooze or seep out. Where a line of springs appears, the term 'spring line' is used. The distribution of springs is related to the nature and relationship of the rocks of an area, together with the profile of the surface relief; they occur at or below the plane where the water table intersects the surface. Where a permeable rock rests on an impermeable one, a spring may appear at the point where the junrtion cuts the hillside. If the layer of saturation is sufficiently extensive, the spring may be permanent, if not, a period of drought may results in an intermittent spring temporarily drying up.

Artesian Wells: There are some wells in which the water automatically flows to the surface. Such wells are called as *artesian wells.* The first well of this type was dug in Artois Province of France and therefore they are called *artesian wells.*

Artesian wells are found in several arid and semi-arid areas of the world, especially where a basin-like structure lies along the side of a hill where rainwater percolates down and collects in adequate quantities in the pervious layer. Examples of this are available in Sahara, Libya, Saudi Arabia, and Iran where a number of oases are found on artesian wells. The most extensive artesian basin is found in Australia. In Australia, there is an area of over 5 lakh square miles which is characterised with aquifer sandstone below the impervious surface layer. This sandstone is exposed on the surface in the Eastern Highlands and gets saturated on account of the heavy rainfall there. In Australia, there are several areas of artesian basins. In some basins the wells are quite deep and go down to depths of one mile or even more.

Geysers: Geyser is a violent ejection of steam and superheated water from an underground source through a hole in the ground. The term is derived from the Icelandic *geysir* (roarer or gusher). The subterranean structure of a geyser comprises a number of water-filled chambers interconnected with a central pipe. The whole system being heated by increasing pressure head due to the height of the column of water. High pressure forces the liquid (water) into a gaseous (steam) phase once a critical temperature threshold has been crossed and the geyser shoots violently into the air, sometimes to a height of 60 metres. Some geysers exhibit a remarkably regular periodicity of ejection, e.g., the Old Faithful, Yellowstone National Park (USA). All known geysers are situated in regions of present or recent volcanic activity, and are themselves an expression of vulcanism. The Waimangu geyser of New Zealand is said to have spouted to a height of more than 400 metres. The temperature of some of the geysers of the Yellowstone Park reaches up to 223°C (400°F).

Deposits of Geyser: The geysers deposit silica, known as *geyserite,* on both inside and outside of the geyser vents. Algae which live in the hot water aid in the precipitation of the geyserite and give rich, beautiful colours to many of the walls of the geyser pools.

The silica occurs in the water in very small quantities, not so abundantly as does the calcium carbonate in the waters of Mammoth Hot Springs. Therefore, in case of geysers, only relatively small deposits of geyserites have accumulated. Some of the deposits are simply spread irregularly over the ground, others are in the form of low mounds, many of which are terraced. Still others have accumulated in irregular columns and small castle-like forms.

Management of Groundwater: Heavy demand for water for irrigation, especially in the developing countries, increased urban population and industrial development have put great stress on groundwater. Rapid withdrawal of water to meet the demand of growing population of teeming millions, agriculture and industries has depleted the underground water table and has impacted the environment at many places. In order to fulfil these needs, vast number of wells using powerful pumps draw great volumes of groundwater to the surface, greatly altering nature's balance of groundwater recharge and discharge.

In dry climates, agriculture is often heavily dependent on irrigation water from pumped wells, especially since major river systems are likely to be already fully utilised for irrigation. Wells are also convenient water sources. They can be drilled within the limits of a given agricultural or industrial property and can provide immediate supplies of water without any need to construct expensive canals or aqueducts.

In earlier times, the small well that supplied the domestic and livestock needs of a home or farmstead was actually dug by hand and sometimes lined with masonry. By contrast, a modern well supplying irrigation and industrial water is drilled by powerful machinery that can bore a hole 40 cm (16 inches) or more in diameter to depths of 300 metres (about 1,000 feet) or more. Drilled wells are sealed off by metal casings that exclude impure near-surface water and prevent clogging of the tube by caving of the walls. Near the lower end of the hole, in the groundwater zone, the casing is perforated to admit the water. The yields of single wells range from as low as a few hundred litres or gallons per day in a domestic well to many millions of litres or gallons per day for large industrial or irrigation wells.

Supply of Water Table

As water is pumped from a well, the level of water in the well drops. At the same time, the surrounding water table is lowered in the shape of a downward-pointing cone, termed the *cone of depression*. The water table depletion often greatly exceeds recharge—

the rate at which infiltrating water moves downward to the saturated zone. In arid regions, much of the groundwater for irrigation is drawn from wells driven into thick sands and gravels. These deposits are often recharged by the seasonal flow of streams that head high in adjacent mountains. Fanning out across the dry lowlands, the streams lose water, which sinks into the sands and gravels and eventually percolates to the water table below. The extraction of groundwater by pumping can greatly exceed this recharge by stream flow, lowering the water table. Deeper wells and more powerful pumps are then required. The result is exhaustion of natural resource that is not renewable except over long periods of time.

There are numerous examples of groundwater depletion in India. In the eastern part of Haryana (Yamunanagar, Kurukshetra, Ambala, Karnal, Panipat, Sonipat, etc.) the water table is increasingly becoming deeper, while in the western districts of Haryana (Bhiwani, Hissar, Sirsa, Jind, Mahem, etc.) it is rising, and at many places, it is only about two metres below the earth surface.

Ground's Significance

Another important environmental effect of excessive groundwater withdrawal is subsidence of the ground surface. Venice (Italy) provides a dramatic example of this side effect. Venice was built in the 11th century AD on low-lying islands in a coastal lagoon, sheltered from the ocean by a barrier beach. Underlying the area are some 1,000 metres (about 3,300 feet) of layers of sand, gravel, clay, and silt, with some layers of peat. Compaction of these soft layers has been going on gradually for centuries under the heavy load of city buildings. However, groundwater withdrawal, which has been greatly accelerated in recent decades, has aggravated the condition.

Many ancient buildings in Venice now rest at lower levels and have suffered severe damage as a result of flooding during winter storms on the adjacent Adriatic Sea. Sea level is normally raised by the effects of coastal storms, and when high tides occur at the same time, water rises even higher. The problem of flooding during storms aggravated by the fact that many of the canals of Venice receive raw sewage, so that the floodwater is contaminated.

Most of the subsidence in recent decades has been attributed to withdrawals of larger amounts of groundwater from industrial well at Porto Marghere, the modern port of Venice, located a few kilometres on the mainland shore. The pumping has now been greatly curtailed, reducing the rate of subsidence to a very small natural rate (about 1 mm per year). However, the threat of flooding and damage to churches and other buildings of great historical value remains. Flood control now depends on the construction of seawalls and floodgates on the barrier beach that lies between Venice and the open ocean.

Perhaps, the most celebrated case of ground subsidence is that affecting Mexico City. Carefully measured ground subsidence has ranged from about 4 to 7 metres (13 to 23 feet). The subsidence has resulted from the withdrawal of groundwater from an aquifer system beneath the city and has caused many serious engineering problems. The volume of clay beds overlying the aquifer has contracted greatly as water has been drained out. To combat the ground subsidence, recharge wells were drilled to inject water into the aquifer. In addition, new water supplies from sources outside the city area were developed to replace local groundwater use.

Contamination of Ground Water

Another major environmental problem related to groundwater withdrawal is contamination of wells by pollutants that infiltrate the ground and reach the water table. Both solid and liquid wastes are responsible. Disposal of solid wastes poses a major environmental problem in the USA, UK, Germany, France, Italy, Netherlands, Denmark, Japan, China and India. The industrial economy of the developed and developing countries provides an endless source of garbage and trash. Traditionally, these waste products were trucked to the town dump and burned there in continually smouldering fires that emitted foul smoke and gases. The partially consumed residual waste was then buried under earth.

In recent decades, a major effort has been made to improve solid-waste disposal methods. One method is high-temperature incineration, but it often leads to air pollution. Another is the sanitary landfill method in which waste is not allowed to burn. Instead, layers of waste are continually buried, usually by sand or clay available on the landfill site. The waste is thus situated in the unsaturated zone. Here it can react with rainwater that infiltrates the ground surface. This water picks up a wide variety of chemical compounds from the waste body and carries them down to the water table.

Once in the water table, the pollutants follow the flow paths of the groundwater. The polluted water may flow towards supply well, which is drawing in groundwater from a large radius. Once the polluted water has reached the well, the water becomes unfit for human consumption. Polluted water may also move towards a nearby valley, causing pollution of the stream flowing there.

In cities like Kolkata, Bhopal, Delhi, Mumbai, Bangalore, Hyderabad, Patna, Kanpur, Agra, Ludhiana, and Chennai, the groundwater has been contaminated at many places, rendering it unfit for drinking purposes. Maintenance of underground water table in a healthy and sustainable condition should be one of the priorities of the planners and administrators. Sooner this wisdom grows, the better for the humanity and the environment.

Kinds of Water

Water that occupies pores, cavities, cracks and other spaces in the crustal rocks is known as *groundwater, subsurface water* or *underground water.* It includes water precipitated from the atmosphere which has percolated through the soil (*meteoric water*), water that has risen from deep magmatic sources, liberated during igneous activity *(magmatic* or *juvenile water)*, and *fossil water* retained in sedimentary rocks since their formation *(connate water)*. Some underground water may also be derived from the percolation of oceanic water, especially in the coastal areas. The presence of groundwater is necessary for virtually all weathering processes to operate. The slow moving underground water can dissolve huge quantities of soluble rocks and carry them away in solution. In some areas, it is the dominant agent of erosion and produces *karst topography,* which is characterised by sink holes, subsidence depressions, collapsed caverns, solution valleys and disappearing streams.

When water reaches the surface of the earth in the form of rain or snow-melt, part runs off down the slopes to join streams through which it ultimately reaches the sea, part is directly evaporated and part is absorbed by vegetation and then mostly returned to the atmosphere. The rest percolates through the surface soil into the bedrock to form the *groundwater* or *phreatic water*. The percolation of water depends on the structure and permeability of the rock. For example, run-off in hot deserts is short-lived, for the rain quickly sinks into sand, or should it fall on some steeply inclined surfaces, run-off may be almost 100 per cent of the precipitation.

Understanding Water Table

The upper level of the zone of groundwater saturation in permeable rocks is known as *water table* or *underground water table*. Above the water table the rocks are unsaturated. It represents the surface at which the pressure in the groundwater is equal to the atmospheric pressure. The elevation of the water table varies seasonally, according to the amount of precipitation, but it also depends on other factors, such as evapotranspiration, and the amount of percolation through the soil. The slope (hydraulic gradient) of the water table is inversely proportional to the permeability of the aquifer. Where the water table reaches the ground surface, it is marked by a spring or by seepage, or artesian well.

Movement of Ground Water

The water moves in the ground is proved by the direct and indirect evidences. For example, wells are pumped dry, yet soon refill with water. Strong flows of water through fractures in rocks are frequently encountered in mine and railroad tunnels. Underground rivers have been discovered in great caverns. Water may move very long distances underground. Except for the water that flows in such openings as fissures or caverns, the movement of ground water is very slow.

Works of Ground Water

Both through its mechanical and chemical work, groundwater profoundly alters the surface of the land. While neither activity is as widespread as are some other physiographic agents, the total results are very great, especially in humid regions where the amount of water in the ground is relatively more. As stated above, groundwater is capable of accomplishing erosion on an enormous scale, but unlike streams, groundwater erodes only by dissolving soluble rocks such as limestone, rock-salt, and gypsum. It then transports the dissolved mineral matter and discharges it into other parts of the hydrologic system or deposits it in the pore spaces within the rock. Groundwater erosion starts with water percolating through joints, faults, and bedding planes and dissolving the soluble rock.

In time, the fractures enlarge to form a subterranean network of caves that can extend for many kilometres. The caves grow larger until ultimately the roof collapses, and a crater-like depression, or sinkhole is produced. Solution activity then enlarges the sinkhole to form a solution valley, which continues to grow until the soluble rock is removed completely.

Through the mechanical process, groundwater accelerates mass movements of land (landslides, rockslides, avalanches, land creep, earth flow, solifluction, etc.). The chemical work of groundwater in extremely varied. Groundwater is an active agent in causing the decay of rocks and thus preparing the land for erosion. The chemical action of water is more pronounced in the limestone (karst) regions.

Important Karst Areas: The word *karst* is a comprehensive term applied to limestone or dolomite areas that possess a topography peculiar to and dependent on underground solution and the diversion of surface water to underground routes. The term comes from the narrow strip of limestone plateau in Croatia and Bosnia-Herzegovina (erstwhile Yugoslavia) and adjacent portions of Italy, bordering the Adriatic Sea.

Significant karst areas are found in different parts of the world. The important areas of karst topography are the Cause region of southern France, Spanish Andalusia, northern Yucatan, Tabasco in Mexico, Jamaica, Puerto Rico, western Cuba, Yunan (China), central New Guinea, Sri Lanka, Myanmar, southwest Celebes, Cherrapunji region (Meghalaya, India), South East Asia, New South Wales, and several states in the eastern and central parts of USA such as Pennsylvania, Maryland, Virginia, Tennessee, Indiana, Florida and Missouri.

In India, the Rohtas Plateau (southwestern Bihar), Sahastradhara (Dehradun), Panchmarhi and Bastar (M.P.), Visakhapatanam (A.P.) and Cherrapunji (Meghalaya) have limestone topography.

Natural Lakes

Lake is an extensive sheet of water enclosed by land, occupying a hollow in earth's surface. It is usually but not necessarily fresh water, from which the sea is excluded.

Lakes vary greatly in area, depth, salinity, eutrophication, altitude, and other characteristics. They may be very small or quite big. Their water may be sweet or saline, though most of the lakes are fresh water lakes. Most of the lakes are situated on the land surface above the sea level, but there are some, such as coastal lagoons, which are at sea level and others, like the Dead Sea and Caspian Sea are below the sea level.

Some of the largest lakes are virtually inland seas (such as Caspian, Superior, Aral, Victoria, Baikal and Balkhash). The Caspian Sea (lake) covers an area of about 4,50,000 sq km and Lake Superior is 560 km (350 miles) in length and 260 km (160 miles) in width with a maximum depth of 400 metres (1,300 feet). Contrary to this, there are numerous small lakes like Anchar, Gangabal, Nagin and Dal (Kashmir), Bhimtal and Nainital (Kumaun Himalayas, India).

The largest man-made lake is Bratsk dammed up in the valley of Angara in Russia, followed by Kariba (Zembezi River). A lake may be seasonal or perennial. Chad is the largest seasonal lake.

It is on the border between Niger and the Republic of Chad which fluctuates in area between 10,000 and 50,000 sq km. The Lake Eyre in south Australia is normally almost wholly dry, its bed consisting of salt crust. Inland drainage lakes which have no outlet to the sea (Sambhar Lake of Rajasthan, India) tend to become increasingly saline. The Dead Sea is a typical saline lake which has a salinity of 238 per cent. The floor of the Dead Sea is more than 800 metres (2,600 feet) below the surface of Mediterranean Sea.

Man-made Lakes

In order to harness water for the generation of hydel power, to provide irrigation water to crops and drinking water to urban places, a number of lakes (water reservoirs) have been constructed by man in almost all the countries of the world. The Ust Ilimsk on a tributary of the Ob River in Russia is one *such* example. The Govindsagar on Sutlej, Mahsamand and Fatehsagar in Udaipur (Rajasthan) are also the examples of man-made lakes.

In the geological history of the earth, lakes are a very temporary and ephemeral feature of the landscape. They may be filled up by river-borne sediments, especially in mountainous areas where heavily laden streams rapidly build lacustrine delta like that of the Jhelum River in the Wular Lake of Kashmir. These sediments eventually fill the lake. Several of the lakes of the Kashmir Valley like Wular, Nagin, Anchar and Dal have shrunk substantially, while Mansabal Lake is only a marshy tract which gets filled with water only at the occurrence of heavy rains.

Similarly, the eastern part of Lake Geneva is being slowly filled in by the Rhone. Various estimates by Swiss and French scientists put the life of the lake at between forty and fifty thousand years if the rates of sedimentation are maintained. The Dal Lake of Kashmir which was about 115 sq km in area during the 16th century, has been reduced to only 16 sq km, and if the present rate of sedimentation continues, the Dal may disappear within the next fifty years, leaving behind a lacustrine plain.

The climatic change, increasing aridity, eutrophication and human encroachments are the other causes leading to shrinkage of fresh water lakes, especially in the densely populated plain areas.

A judicious use of lakes is imperative to conserve them and to maintain the lake ecosystems in sustainable condition. It is all the more important because the fresh water supply for drinking purposes to many of the cities and towns of the world exclusively depends on lakes.

Lakes Formed by Landslides

Most of the lakes have their origin because of the endogenic or exogenic forces. Some of the river beds are dammed by landslides and earth movements which result in the formation of lakes. In the hilly areas, especially in the weak stratas and breccia rocks at the occurrence of heavy rains, there occur massive landslides. The heavy landslides may block the river channel, and the portion above the landslide dam is converted into a dam. Lakes of this type are usually temporary as the water level above the dam keeps on rising and soon overtops it or the dam ruptures under the impact of water. Several examples of such short-lived lakes are available from the Himalayas. For example, in 1840, a lake about 64 kilometres long and 300 metres deep was formed in the *Nangaparbat* area as a result of damming of the Indus River. The barrier lasted for nearly six months. It broke in June 1841 and the lake was drained of its water within 24 hours. Across the Ganga River, the Gohana Lake was created in 1893 due to massive landslide. The lake lasted for about eight months and resulted into devastating flood when the dam was ruptured. The Dong-Hu Lake near Wuhan city of China is also an interesting example of natural-dammed lake.

Lakes of Tectonic Origin

Folding and Faulting: Folding and faulting produce hollows in the earth's crust. The hollows thus formed may-contain either salt or fresh water. Lakes Victoria, Tanganyika, Nyasa, Albert, Rudolf in East Africa, Titicaca on the high interment Andean plateau in South America, Caspian Sea, Aral Sea, Baikal, and Balkhash in Asia are some of the major examples of tectonic lakes. The Great Salt Lake (USA), is also an example of such lakes. The Lake Baikal in Central Asia is defined by a whole series of faults. It is the deepest lake in the world. Its area is about 11,580 square miles. The Tarim basin, enclosed by the inward facing fault-line scarps of the Altyn Tagh on the south and Tien-Shan on the north, once occupied by a vast lake, is now covered with sands, gravels, and the Lob-Nor marshes.

Rift Valleys: A linear depression or trough created by the sinking of the intermediate crustal rocks between two or more parallel strike slip faults is known as a *rift valley.* The Jordan-East African rift valley is a major example, containing the Dead Sea, Lake Tanganyika, Lake Nyasa (Malawi) and a series of smaller lakes. The *Dead Sea* occupies the deepest part of the northern section of this trough; its surface is about 394 metres (1,294 feet) below the sea level and its greatest depth is 400 metres (1,300 feet). It exemplifies the elongated character of a *rift-valley lake,* for it is 88 kilometres (55 miles) in length and only 16 kilometres (about 10 miles) in width.

Volcanic Lakes: The funnel-shaped basin, surrounding the vent at the summit of a volcano or on its flanks, is known as a *crater.* When the crater of an extinct or dormant volcano is filled with water, it is known as a *crater lake.* The *crater lakes* are generally circular in shape. The Danau Lake of Sumatra (Indonesia) is an interesting example of a crater lake. This is the *largest crater lake* in the world, which lies in a caldera 1,900 sq km (750 square miles) in area among the Batak Highlands of northern Sumatra. Steep walls 600 metres (2,000 feet) high encircle the lake, except for its river outlet in the southeast, where the Soengai Aasahan plunges through a series of gorges and over a waterfall 135 metres (443 feet) high.

Occasionally, a lava flow may block a valley and so form a lake basin. One such has made a barrier across the Jordan Valley, damming up the Sea of Galilee and another in the East African Rift-Valley is responsible for Lake Kivu. Lake Van in eastern Turkey, at a height of about 1,700 metres (5,600 feet) above the sea level, was formed by outpouring of lava from the large volcano of Nemrut with a caldera 10 kilometres (6 miles) in diameter, blocking the valley of a head stream of the Euphrates, so forming a lake. A striking example in Japan is Lake Chuzenji in Nikko National Park, whose overflow spills out over a lava barrier to form the 100 metres (330 feet) Kegon Falls.

Lakes: Depositional

Lakes which come into existence because of the depositional action of running water, wind, glacier, sea waves and underground water are known as the *depositional lakes.* Lakes ponded up by deposition are known as *barrier lakes,* since their waters are contained, in part at least, by some natural dam, though often of a very short-lived nature. A landslide or avalanche may block a valley, so that a river course is checked and the water ponded up; such a dam is usually unstable, and when pressure of water ultimately sweeps it aside, disastrous floods may surge down the valley.

Ox-bow (cut-offs) lake is a crescent-shaped lake occurring on a river floodplain, having once been part of a river (meander) that has been cut through by the depositional work of river. The ox-bow lake is also known as a *bayou, billabong* or *mortlake.* There are numerous ox-bow lakes in the middle and lower parts of the Ganga plain, Sunderban delta, Brahmaputra and Indus valley and the Mississippi delta. Brackish coastal lakes are enclosed by sand bars built up by coastal depositions, such as the Haffe of the Baltic Sea, Lake Pulicat (Tamil Nadu, India), and Lake Chilka on the coast of Orissa (India). The main condition necessary for the development of sea coast depositional lakes is an ample longshore drift of material, together with an irregular coastline which the sea is smoothing off by building up these bars of sediments.

Morainic Deposits: Morainic deposits are accumulation of heterogeneous rubbly material, including angular block of rock, boulders, pebbles and clays that has been transported and deposited by a glacier or ice-sheet. In undulating lowland country, covered with uneven deposits of glacial debris, thousands of oddly shaped lakes, ranging from pools to extensive sheets, may be found. The lake plateaus of Mecklenburg in Germany, and of Pomerania of Poland are striking examples. A special variety of lake due to glacial deposition is known as a *kettle-hole,* where a large ice block was buried in drift, and so left upon melting a depression which now contains water. Numerous such examples occur in the Kettle Moraine country of Wisconsin, from which the name was derived.

Ice Dams: An obstruction caused by ice floes when they accumulate during a melting phase, either in a river or in a lake, may lead to ice jam, which develops an *ice-dammed lake.* The ice barrier, formed by a glacier or an ice sheet which causes the ponding of meltwaters or natural drainage, may also create an ice-marginal lake. The Marjelen Lake, formed by the Aletsch glacier in Switzerland, is an example of the ice-dammed lake. At the margin of Greenland ice cap numerous fjords are glacier-dammed to form ribbon lakes 15-30 kilometres (10-20 miles) in length. Over fifty named ice-dammed lakes are concentrated along the coastal strip of southern Alaska, most within 80 kilometres (50 miles) of the ocean. The Vatnsdalur Lake of Iceland is also an example of ice-dammed lake.

Vegetation Dams: The blocking of slowly flowing streams near their mouths form vegetation dam which lead to the development of small depositional lakes. Such are the Gooren and Vennen lakes in the southern Netherlands and northeastern Belgium, respectively.

Calcareous Dams: Some unusual deposition lakes in the erstwhile Yugoslavian Karst have been formed by the growth of barrier of calcareous material across a river. Lake Plitvice in central Yugoslavia is an example of calcareous-dam lake.

Lakes: Erosionable

Glaciation: Glacier is one of the important agents of erosion. The erosive process of glaciers and ice-sheets creating cirques, U-shaped valleys, and the irregular surface of a glaciated lowland, will provide depressions in which water can accumulate. Lakes are one of the most characteristic features of a formerly glaciated landscape. Much of the charm of Scandinavian countries (Norway, Sweden, Finland, Denmark), Great Bear Lake, Great Slave Lake, Lake Winnipeg (Canada), Lake District of England, North Wales, Italian Alps, Himalayas (Mansarovar, Rakastal, Gangabal, etc.) is because of the presence of glaciated lakes. One of the most striking groups of 'Finger Lakes' is in New York State. The Finger Lakes add much to the natural beauty of a region in which resorts and state parks attract many tourists. The region is the centre of New York State's wine industry. The low hill country in which they lie once drained south to the Susquehanna River, but ice-lobs scoured deep valleys in which the lakes now lie and breached the preglacial divide, so that they now drain northward, ultimately to Lake Ontario. The lake plateau of Finland and parts of the Canadian Shield are lake-studded.

A glacial lake may lie in a U-shaped valley, the lower part of which is gouged into solid rock, while the morainic dam across the mouth of the valley increases both the depth and area of the water.

A large number of lakes is to be found in Scandinavia and Finland (Saimaa L.) across the boundary of ancient shields, known as the *glint-line.* They are partly due to erosion, both by rivers and glaciers, and partly due to the blocking of these eroded valleys by moraines. In Scandinavia, a series of these long, narrow lakes is strung out along the valley of each stream flowing in more or less parallel courses to the Baltic Sea. The highest glacial lake of India is *Devtal* which is situated at a height of about 5,460 metres (17,745 feet) above the sea level in Garhwal Himalayas.

Solution Lakes: The removal of certain rocks in solution may produce hollows which can contain small lakes. Some of the *meres* (flashes) of Cheshire are probably due to local subsidence as a result of removal in solution of underlying beds of rock-salt. Lough Derg in western Ireland is a large, shallow water area, which has developed because of dissolution of limestone rocks. Solution often forms great underground caverns, which near the base of the limestone beds may contain subterranean lakes. The collapse of the limestone roof may produce a long, narrow surface lake. In the limestone region of the erstwhile Yugoslavian Karst, many lakes occur on the floor of the *Polia* which are due, in part at least, to solution. Their floors may carry seasonal lakes, most of which degenerate into salt-marsh or disappear altogether in summer. A few, however, like Lake Skadar on the Yugoslav-Albanian frontier and Lake Plitvice (Yugoslavia) are perennial. The Nainital Lake in Kumaun Himalayas (Uttaranchal, India) is also considered as a solution lake, though faulting has also played a role in its formation.

Wind: The defiating action of the wind in desert region can produce great hollows reaching ground water. Such hollows are quite numerous in the limestone plateau west of the Nile valley. The largest is the Qattara (Egypt), the floor of which reaches

134 metres (436 feet) below the sea level. Similar though shallower depressions are to be found in the Kalahari desert, West Australian deserts, and in Mongolia. Many contain oasis or salt lakes, indicating that the water table, the limiting factor in the wind's deflating effect has been reached.

The Qattara depression in Egypt, and the Shott-el-Jerid and Shott-Melrhir on either side of the Algerian-Tunisian frontier are the typical examples of lakes which came into existence because of the erosional action of wind.

The Division

Lakes may be classified according to the mode of origin of the hollows which contain their waters. The most important categories are produced by erosion, deposition, earth movements and volcanic activity. Some of the lakes as stated above are man-made. There are, however, some lakes which came into existence because of more than one reason working in conjunction. For example, a sheet of water can accumulate in an erosion hollow or in natural barriers caused by deposition. On the basis of mode of origin, lakes may be grouped under the following categories: a) Erosional lakes; b) Depositional lakes; c) Lakes of tectonic origin; d) Volcanic lakes; e) Lakes formed by landslides; and f) Artificial (man-made) lakes.

Rivers in Nature

Rivers have often been described as the lifeblood of the earth. They redistribute mineral nutrients important for soil formation and plant growth and serve society in many ways. Not only do rivers provide us with essential water supplies, but they also receive, dilute and transport wastes, provide critical cooling water for industry and form one of the world's most important transportation networks. Rivers have been of fundamental importance throughout human history.

Intersecting Profile

The bank-to-bank profile across the width of a river is known as *cross profile.* In other words, a cross profile is a diagrammatic means of depicting the transverse profile of a valley in order to demonstrate the slope gradients, the terrace sequence, etc. It gives an idea about the width of the valley and its varying depth. As stated above, two kinds of processes are involved in the development of a river valley: i) valley deepening, and ii) valley widening. In the initial stage of valley development, the river cuts mainly downward, i.e., the river deepens its bed by vertical corrosion or down cutting. Where this process is dominant, the valley is deep and narrow like a gorge. But, under normal climatic conditions, after a period of time, the rocks on either banks of the river start breaking by weathering, gullying and slipping and then the mouth of the gorge starts opening up and the valley of the river becomes V-shaped.

In this stage, the amount of water and the load of river are both small and tributary streams are either absent or in their formative stage. With the passage of time the down cutting becomes slow, but the processes of corrosion and valley widening continue and become more important. Consequently, the width of the valley goes on increasing and in mature stage the river valley becomes wide and relatively shallow.

In the mature stage, there is more lateral erosion than the downward cutting. Where the river bends, the curve of the bend tends to increase by lateral erosion. In this situation, the currents are faster on the convex side of the valley. Consequently, the river cuts its convex (steep) slope more than the concave (gentle) slope. Thus, by the process of differential erosion on two sides of the valley, in due course of time, a chain of meanders with interlocking spurs projecting into the valley come into existence. With the passage of time, the size of the meanders increases and the lateral cliffs also start getting eroded. In this way, the floodplain widens and ultimately becomes wider than the meander belt. The floodplain built by this process is an erosional feature.

Longitudinal Profile: The profile of a river from source to mouth is known as a *longitudinal profile*. In other words, a longitudinal section of the course of a river drawn along the river. In German language, the longitudinal profile of a river is known as *Thalweg*. Near the source of a river, the erosive power of the river is relatively small, because both the volume of water and the load in the river are small. Generally, the maximum erosion takes place in the region mid-way between the source and the mouth.

In the upper reaches of a river irregularity of surface are found in the river bed resulting in the formation of rapids, cataracts, and waterfalls. These irregularities may be either due to the original irregularities of the surface or they may be caused by differential erosion of rocks of different hardness in the bed of the river. But, these irregularities in the bed of the river are temporary, because as the erosion proceeds, the irregularities are destroyed and levelled down. Striking examples of features caused by such differential erosion of hard and soft rocks are provided by rapids and waterfalls. The waterfall recedes backward.

In the middle and lower parts of the river, the profile is more smooth. The river channel gets divided and oscillates between the levees. In the lower part one may observe the braided channels, abundance of meanders and ox-bow lakes.

Forms of Land

Land Forms in the Upper Course of a Graded Stream: The average balanced state of operation of a stream is known as the *grade of the stream.* In fact, it is an equilibrium condition of a stream which is referred as a *graded stream.* A graded stream can discharge not only the surplus run-off produced by the basin, but also the solid load that the tributary channels supply. This transport capability is related to the gradient of the channel. In general, steeper the gradient, the higher the stream velocity and the greater the ability of the stream to carry sediments. The major types of *waterfalls* are: i) vertical barrier falls, ii) plateau falls, iii) fault-plane falls, and iv) hanging valley falls.

Vertical Barrier Falls: The areas of sedimentary rocks which are intruded by hard igneous rocks in the form of vertical dykes are the ideal places for the development of *vertical barrier waterfalls.* In such a situation, the sedimentary rock erodes easily, while the igneous dykes endure erosion. In such cases, the resistant dykewall stands in the way of the recession which is consequently more stationary. In the Yellowstone Park (USA), the fall on Yellowstone River is an example of a vertical barrier fall.

Plateau Falls: Plateau falls are found in areas where rivers descend suddenly from the plateaus of igneous rocks to the lowlying, softer rocks of the coastal plains. The Livingstone falls of the Congo River belong to this category. The Jog falls in the Western Ghats (Karnataka) are also a good example of plateau falls.

Fault Plane Falls: When the upper part of a river is uplifted by earth movement and fault develops in its course, then the river forms a waterfall along the fault scrap because of the sudden change in level. The Hundroo falls on the Subarnarekha River near Ranchi (India) is a good example of fault-plane falls.

Hanging Valley Falls: The hanging valley falls are generally found in the glaciated mountains. They are also found along the sea coast under the influence of wave action. On the rivers of Chhotanagpur plateau there are numerous hanging valley falls. The Parkchak waterfall (Laddakh), the Thajwas fall (Sonmarg, Kashmir) and numerous waterfalls of Norway, Canada and Alaska belong to this category.

Land Forms in the Middle Course of a Graded Stream: In the middle part of its course a river develops mature features, and the valley becomes wider. After crossing the mountainous and hilly regions, the river debouches into the plain. In the plain area the slope is gentle, owing to which the river is slackened. The slackened river is unable to carry the large-sized shingles, pebbles, gravel, and sand. Consequently, they start being deposited. This deposit is triangular in shape through which the river flows in several braided channels. This gently sloping accumulation of coarse alluvium deposited by a braided stream is known as *alluvial fan* or *alluvial cone*. In some cases, the neighbouring alluvial fans may meet and coalesce with each other as they develop and form an extensive alluvial plain in the foothill or piedmont zone. This type of land form is known as *piedmont alluvial plain*. The alluvial fans are built mostly during the rainy season, when the volume of water as well as the sediments are enormous.

Bajada (Spanish, *Bahada)* is a land form with gentle and sloping surface leading down from a mountain front to an inland basin, in an arid or semi-arid basin. It is composed of unconsolidated materials, such as sand, gravel and angular screen which together mantle the underlying piedmont.

In the middle course of a river, the valley becomes wider through the weathering of sides. As the water flows round a bend, the river tends to accentuate the curve. Since the current impinges most strongly on the convex side (outside of the curve), the maximum erosion, even undercutting, takes place there. There is a little erosion, and even some deposition, in the slack of the current on the inside of the bend. Thus, initial 'swing' in the river may be transferred into a meander. The Ganga River makes numerous meanders between Haridwar (Uttaranchal) and the Sundarban delta region in West Bengal.

Another important land form of the middle course of a graded river is the development of *ox-bow lakes.* The ox-bow is a crescent-shaped lake occurring on a river floodplain, having once been part of a river (meander) that has been cut through the abandoned by lateral erosion of the banks at the meander neck. The ox-bow lake is also known as *billabong, mortlake* or *bayou.*

Floodplain is also a characteristic land form of the middle course of a river. It is a flat tract of land bordering a river, mainly in its middle and lower reaches, and consisting of alluvium deposited by the river. The sides of a floodplain may be several kilometres apart. In time of flood, when the river overflows its banks, sediment is deposited along the banks and in the channel itself. The process elevates the river channel above the level of the floodplain. The raised banks are known as *levees,* and the lower land behind them as *back swamps.* The raised banks or levees help in the development of a *yazoo* stream or river tributary.

A *river terrace* is a portion of the former floodplain of a river now abandoned and left at a higher level as the stream down cuts, usually from rejuvenation. The term is used to describe both the bench-like form and the alluvial deposits of the former floodplain. The river terraces generally develop in the upper and the middle course of a river.

Land Forms in the Lower Valley: In the lower part the speed of the river is further reduced. The river wanders or shuffles in a series of sweeping meanders over a broad and almost levelled valley. The channel of the river becomes shallow and wide. At the occurrence of heavy rains the river water overflows the banks. Relics of former meanders are left in the form of '*ox-bow lakes*', or '*cut offs*'.

In the lower part of a river's course deposition becomes of major importance, and vertical corrosion ceases except for the incising of the channel in the floor of the floodplain as the meanders move downstream.

Many large, heavily laden rivers, especially when they flow from an extensive mountainous area, spread out sheets of material which may split the stream into complicated channels. Such rivers are said 'braided'.

Delta: In the lower course of a graded river, delta is also a land form. It is a deposit of sediments formed at the mouth of a river where it enters a lake or sea. It is normally built up only where there is no tidal or current action capable of removing the sediment as fast as it is deposited, and hence the delta builds forward from the coastline. This process of building up is complex, and leads to the formation of a number of separate channels, isolated lagoons, levees, marshy grounds and a network of small creeks.

In general, deltas are triangular in shape, but on the basis of shape they may be classified into the following five categories: i) arcuate delta (fan-shaped), ii) bird's foot (digitate) delta, iii) estuarine delta, iv) cuspate or tooth-shaped delta, and v) lacustrine delta.

Arcuate Delta: The delta in which the outermost margin exhibits an arc-like form, convex towards the sea, is known as *arcuate delta*. The delta of Nile River is the typical example of arcuate delta. The deltas of the Ganga, Hwang-Ho, Po and Rhine rivers are also good examples of arcuate delta.

Bird's Foot Delta: This type of delta is formed by the outgrowth of natural river levees into a body of a water to form a finger-like pattern, reflecting the number of distribuτary streams. Its best example is the delta of Mississippi River.

Estuarine Delta: Where the river enters the sea through a single mouth or estuary a long and narrow delta is formed. Normally, the estuary is open and the river sediments are removed by the waves and currents. But, when the mouth of the river is submerged below the sea, a narrow linear delta is formed. The deltas of the Amazon and the Congo rivers are the typical examples of estuarine delta.

Cuspate Delta or Tooth-Shaped Delta: This is a symmetrical delta, usually formed where a river debouches in a straight coastline, and in which the sedimentary material is deposited evenly on either side of the river mouth. The delta of Tiber river (Italy) is an example of the cuspate delta.

Lacustrine Delta: A delta built into lake by a heavily laden stream is known as a *lacustrine delta.* The fine silt deposited in the lake basin ultimately result to fill up the lake. The biggest lacustrine deltas are those which are being built out into the Caspian Sea by the Volga, Ural and Kura rivers.

System of Drainage

A river and all its tributaries within a single drainage basin is known as the *drainage system.*

Equilibrium in River System: A river system works as a unified whole. Any change in one part of the system affects the other parts. The major factors that determine stream flow (discharge, velocity, channel shape, gradient and load) constantly change towards a balance, or equilibrium, so that eventually the gradient of the stream is adjusted to accommodate the volume of water available, the channel's characteristics, and the velocity necessary to transport the sediment load.

River system is dynamic, it changes over space and time. One of the important characteristics of a river system is that it functions as a unified whole. A river is in equilibrium if its channel form and gradient are balanced so that neither erosion nor deposition occurs. Rivers are constantly adjusting to approach this ideal condition. This adjustment is important in understanding the natural evolution of the landscape.

Graded Stream: A stream that has attained a state of equilibrium or balance between erosion and deposition so that the velocity of the water is just great enough to transport the sediment load supplied from the drainage basin and neither erosion nor deposition occur.

To summarise, the basic fluvial system is a drainage basin – an open system. Streams produce fluvial erosion, which supplies weathered and wasted sediments to transport to new locations, where they are laid down in a process known as *deposition.* A stream is a mixture of water and solids, carried in solution, suspension and by mechanical transport. Alluvium is the general term for the clay, silt and sand transported by running water.

Stream transport material occurs as dissolved load in solution, as suspended load held aloft in the stream and as bed load dragged along the stream bed by traction, or rolled and bounced along the saltation.

All the streams have a gradient and work to establish a graded profile over distance, with interruptions triggering adjustments. Slope is the critical factor in a graded stream's maintenance of an equilibrium condition between water and solid materials. Stream form, and operation result from complex interactions of slopes, discharge, load and channel characteristics, all variable within different climates and with different rock types. These functional considerations are embodied in the dynamic equilibrium approach to understanding the fluvial landscape.

Various landforms are associated with the action of flowing water. Waterfalls, gorges, V-shaped valleys, river terraces, floodplains, levees, meanders, ox-bow lakes, and deltas are some of the major erosional and depositional features of running water (rivers).

Development of River Valley

Valley is a linear depression sloping down towards a lake, sea or inland depression. Valleys are present over most of the land surface of the earth. They are known by such names as *gully, draw, ravine, gulch, hollow, run, arroyo, gorge* and *canyon.* All the valleys were cut by running water. There are some *diastrophic valleys* also like the Death Valley, the Great Valley of California, the Vale of Chile and the Jordan Valley. There are some valleys in valley which are found in the rejuvenated topography.

A valley takes form through the operation of three concomitant processes: i) valley deepening, ii) valley widening, and iii) valley lengthening.

The development of a valley, its deepening and widening are influenced by several processes. They are: i) hydraulic action,

ii) corrosion on the floor of the valley, iii) pothole drilling along the valley floor, and iv) corrosion and weathering of the stream bed. The development of a river valley may be examined under two heads:

1. Cross profile, and
2. Longitudinal profile.

Process of Flow

Stream-related processes are termed as *fluvial* (from the Latin *fluvius,* meaning 'river'). The landforms, shaped by running water, are called as *fluvial landforms.* Geographers seek to describe recognisable stream patterns and the fluvial processes that created them. Fluvial streams, like all natural systems, have characteristic processes and produce predictable landforms. Yet, a stream can behave with randomness, unpredictability and disorder.

Insolation and gravity are the driving forces of fluvial systems because they power the hydrologic cycle. Individual streams vary greatly, depending on the climate in which they operate, the variety of surface composition and topography over which they flow, the nature of vegetation and plants cover, and the length of time they have been functioning in specific settings.

Weathering, wind and ice dislodge, dissolve or remove surface material in the process called *erosion.* Thus, streams produce fluvial erosion, which supplies weathered sediment for transport to new locations, where it is laid down in a process known as *deposition.* A *stream* is a mixture of water and solids, carried in solution, suspension and by mechanical transport. *Alluvium* is a general term for the clay silt transported by running water.

It is interesting to note that running water of the hydrologic system of the earth is the most important agent of erosion. Moreover, stream valleys are the most abundant and widespread land forms on the continents.

The work of streams modifies the landscape in many ways. Land forms are produced by erosive action of flowing water and the deposition of stream-transported materials. Rivers create floodplains as they overflow during floods.

Largest Rivers on Earth Ranked by Discharge Volume

Rank by Volume	*Average discharge at mouth in thousands of cms (cfs)*	*River (with tributaries)*	*Outflow/ location*	*Length km (mi)*	*Rank by length*
1	212.5 (7500)	Amazon (Ucayali, Tamboo, Ene, Apurimac)	Atlantic Ocean/ Amapa-Para, Brazil	6570 (4080)	2
2	79.3 (2800)	La Plata (Parana)	Atlantic Ocean/ Argentina	3945 (2450)	16
3	39.7 (1400)	Congo, also known as the Zaire (Lualaba)	Atlantic Ocean/ Angola, Zaire	4630 (2880)	10
4	38.5 (1360)	Ganga (Brahmaputra)	Bay of Bengal/ Indian Ocean	2898 (1800)	23
5	21.8 (770)	Yangtze (Chang Chiang)	East China Sea/ Kiangsu, China	5960 (3720)	4
6	17.4 (614)	Yenisey (Angara, Selanga or Selenge, Ider)	Yenisey Gulf of Kara Sea/Siberia	5870 (3650)	5
7	17.3 (611)	Mississippi (Missouri, Ohio, Tennessee, Jefferson, Beaver head, Red Rock)	Gulf of Mexico/ Lousiana	6020 (3740)	3

Contd...

Rank by Volun	*Average discharge at mouth in thousands of cms (cfs)*	*River (with tributaries)*	*Outflow/ location*	*Length km (mi)*	*Rank by length*
8	17.0 (600)	Orinoco	Atlantic Ocean/ Venezuela	2737 (1700)	27
9	15.5 (547)	Lena	Laptev Sea/ Siberia	4400 (2730)	11
10	14.2 (500)	St. Lawrence	Gulf of St. Lawrence/ Canada and U.S.	3060 (1900)	21
33	2.83 (100)	Nile (Kagera, Rluvuvu, Luvironza)	Mediterranean Sea/Egypt	6690 (4100)	1

Basin of Drainage

An integrated system of tributaries and a trunk stream, which collect and funnel surface water to the sea, a lake or some other body of water, is known as a *river system.* A drainage basin is a synonym of river system. *Drainage basin* is the basic spatial geomorphic unit of a river system, distinguished from a neighbouring basin by ridges and highlands that form divides. A drainage basin is bounded by a divide (ridge) beyond which water is drained by another system.

There are three main processes involved in a river system: i) erosion, ii) transportation, and iii) deposition.

Erosion: The processes that loosen sediment and move it from one place to another on the earth's surface is known as *erosion.* Running water is a powerful land forming agent. The action of water on the earth's surface can be observed everywhere. Flowing as a sheet across a land surface, running water picks up silt particles and carry them down slope into ruts, gullies and stream channels. When rainfall is heavy, stream and rivers swell, lifting large volumes of sediment and carrying them downstream. In this way, running water erodes mountains, hills and plateaus, carves valleys and deposits sediments.

Running water erodes the rocks either by the chemical or the mechanical method. Some of the rocks over which the water flows (limestone) are dissolved in water. This process of dissolution is known as *chemical erosion.* The mechanical erosion is caused by loose solid material carried down by river, in wearing away fragments from its bed and its banks. Corrasion of the bed, thereby causing a deepening of the channel is called *vertical erosion,* while corrasion of the banks is called *lateral erosion.* Most of the corrasion effected by a river, however, is due to pebbles, gravel and sand that it drives along its banks. Thus mechanical erosion is more important than the chemical erosion. The rock fragments also collide against one another and in the process get disintegrated into small, smooth and rounded pieces of rock and ultimately into sand and silt. The process by which there is a progressive downstream reduction in the size of the materials constituting the load of a river by constant collision is known as *attrition.* In brief, *stream erosion* is the progressive removal of mineral material from the floor and sides of the channel, whether bedrock or regolith. *Stream transportation* consists of movement of the eroded particles dragged over the stream bed, suspended in the body of the stream, held in solution in ions. *Stream deposition* is the accumulation of transported particles on the stream bed and floodplain, or on the floor of a standing body of water into which the stream empties. Erosion cannot occur without some transportation taking place, and the transported particles must eventually come to rest. Thus, erosion, transportation and deposition are the three phases of erosion.

Transportation: The solid matter carried by a stream is the stream load. It is carried in three forms: Dissolved matter is transported invisibly in the form of chemical ions. The running water and all streams carry some dissolved material. Sand, gravel, pebbles, shingles, and larger particles move as bed load close to the channel floor by rolling or sliding. Clay and silt are carried by suspension, that is, they are held up in water by the upward elements of flow in turbulent eddies in the stream. This fraction of the transported matter is the suspended load. Of the three forms, suspended load is generally the largest. The larger rivers like the Amazon, La Plata, Congo, Yangtze, Mississippi, Nile, Indus, Brahmaputra, Ganga, Lena, St. Lawrence, etc., carry over 90 per cent of their loads in suspension.

In general, the more the velocity of a river, the more load it can carry, and more the load, the greater is the corrasive power of the river. But after a certain point, corrasion decreases as the load increases, and the increase in load reduces the velocity of the stream. As increase in the velocity of a river results in an increase in its corrosive power as well as transporting capacity, the river is able to perform several times its normal work of erosion and transportation during the short period of flood in the rainy season.

Deposition: Accumulation of transported particles on a stream bed, upon the adjacent floodplain or in a body of standing water is known as *stream deposition.* As the velocity of a river decreases, it begins to deposit its load of rock fragments, sand, and silt. When a fast flowing hill stream debouches into the plain, there is deposition of stony debris, pebbles, shingles, sand and silt. These depositions lead to the formation of the alluvial cones or alluvial fans. When the river reaches the more level surface of its valley, its flow slackens further, and the river starts depositing sand and silt in its bed resulting in the formation of *diaras.* At the times of flood, the river inundates its adjacent areas and deposits sand and silt outside the confines of its channel. The maximum deposition takes place in the delta region when the river enters the sea or lake. In the delta region the flow becomes very slow, and there is large-scale deposition of sand and silt near the mouth of the river. Because of the presence of salt in the sea water, the sediments brought by the river do not easily mix in the sea water and get precipitated and deposited near the river's mouth. This results in the formation of deltas.

The Features

A stream does not occur as a separate, independent entity. Every stream, river, gully and ravine are part of a drainage system, with each tributary intimately related to the stream into which it flows and to the streams that flow into it. Every stream has tributaries, and every tributary has smaller tributaries, extending down to the smallest gully.

The characteristics of a river change from the source to the mouth (sea). Some of the more important relationships and generalisations of a river in the upper and lower courses are as follows:

- The number of stream segments (tributaries) decreases downstream in a mathematical progression.
- The length of tributaries becomes progressively greater downstream.
- The gradient, or slope, or tributaries decrease exponentially downstream.
- The stream channels become progressively deeper and wider downstream.
- The size of the valley is proportional to the size of the stream and increases downstream.

These relationships contribute the basis for the conclusion that streams erode the valleys through which they flow.

The Differences

Rivers are highly complex systems influenced by a number of variables and, as in the case with so many natural systems, if one variable is changed it produces a change in others. The most important variables are: i) discharge, ii) velocity, iii) gradient, iv) sediment load, and v) base level. A brief account of these variables has been given below:

Discharge: Discharge is the amount of water passing a given point during a specific interval of time. It is usually measured in cubic metres per second.

Velocity: The velocity of flowing water is not uniform throughout the stream channel. It depends on the shape and roughness of the channel and on the stream pattern. The velocity usually is greatest near the centre of the channel and above the deepest part, away from the frictional drag of the channel walls and floor. As the channel curves, however, the zone of maximum velocity shifts to the outside of the bend, and a zone of minimum velocity forms on the inside of the curve.

Steep gradients produce rapid flow, which commonly occurs in high mountain streams. Where slopes are very steep, waterfall, rapids and cataracts develop, and the velocity approaches that of free fall. Low gradients produce slow, sluggish flow, and where a stream enters a lake or an ocean, its velocity is soon reduced to zero. The velocity of flowing water in a given channel also depends on water volume. The greater the volume, the faster the flow.

Stream Gradient (Slope): Stream gradient or slope is also one of the important factors that controls the stream flow. The gradient of a stream is steepest in the headwaters and decreases downslope. The longitudinal profile (a cross section of a stream from its headwaters to mouth) is smooth, concave, upward curve that becomes very flat at the lower end of the stream. The gradient is usually expressed in the number of metres the stream descends for each km of flow.

Sediment Load: The amount of sediment carried by a stream at a given time is known as *load of stream*. Flowing water in natural streams provides a fluid medium by which loose, disaggregated regolith is picked up and transported to the ocean, sea or lake. The capacity of a stream to transport sediment increases to a third or fourth of its velocity, i.e., if the velocity is doubled, the stream can move from 8 to 16 times as much sediment. Running water is the major agent of erosion, not only because it can abrade and erode its channel, but because of its enormous power to transport loose sediment produced by weathering. Within a stream system, sediment is transported by three ways: i) Fine particles are moved in suspension (suspended load); ii) Coarse particles are moved by traction, rolling, sliding, and siltation along the streambed (bed load); and iii) Dissolved material is carried in solution (dissolved load).

Base Level: The base level of a stream is the lowest level to which the stream can erode its channel. It is a key feature in the study of stream activity. The base level is, in effect, the elevation of the stream's mouth, where the stream enters an ocean, a lake, or another stream. A tributary cannot erode lower than the level of the stream into which it flows. Similarly, a lake controls the level of the erosion for the entire course of the river that drains into it. The levels of tributary junctions and lakes are temporary base levels: lakes can be filled with sediments, or drained, and streams can be established across the former lake bed. For all practical purposes, the ultimate base level is the sea level, because the energy of a river is quickly reduced to zero as it enters ocean. Exceptions are fault valleys like Death Valley, which is faulted below sea level. Even sea level can change, however, and as base level or sea level changes, the longitudinal profile of the river changes and the streams adjust to the new conditions.

Bibliography

Ahmad, A.: *Physical Geography*, Rawat Publications, Jaipur, 1993.

Andrew, S. and Antony, O.: *The Physical Geography of Africa*, Oxford University Press, New York, 1996.

Bacon, R. S. and Rubenstein, J. M.: *The Cultural Landscape: An Introduction to Human Geography*, Prentice Hall, New Delhi, 1990.

Bambrick, S.: *The Cambridge Encyclopedia of Physical Geography*, Cambridge University Press, New York, 1994.

Basham, A. L.: *A Handbook of Physical Geography*, Grove Press, New York, 1959.

Beaumont, P.: *The Middle East: A Geographical Study*, John Wiley and Sons, New York, 1988.

Bhatia, B. M.: *Poverty, Agriculture and Economic Growth*, Vikas Publishing House, New Delhi, 1977.

Billihgs, W. D.: *Plant, Man and the Ecosystem*, MacMillan, London, 1971.

Brun, S. D.: *Cities of the World: World Regional Urban Development*, Harper and Row, New York, 1983.

Burnham, C. P.: *Land Evaluation*, Oxford University Press, New York, 1981.

Candolle, Alphonse De: *Origin of Cultivated Plants*, Hafner, New York, 1886.

Carlson, L.: *Geography and World Politics*, Prentice Hall, Englewood, 1958.

Carvalho, C. M.: *Readings in Cultural Geography*, Chicago University Press, Chicago, 1962.

Chandna, R. C: *A Geography of Population*, Kalyani Publishers, Ludhiana, 1992.

Chapman, G. P. and K.M. Baker: *The Changing Geography of Asia*, Routledge, New York, 1992.

Chatterjee, S. P.: *Rural Settlement and Land Use: An Essay in Location*, Hutchinson, London, 1966.

Christaller, Walter: *Central Places in Southern Germany*, Prentice Hall, New Jersey, 1966.

Clark, W. A.: *Migration and Development*, The Hague, Mouton, 1975.

Clarke, J. I.: *Geography and Population: Approaches and Applications*, Pergamon Press, New York, 1984.

——————: *Population Geography*, Pergamon Press, Oxford, 1972.

Claval, P.: *Introduction to Regional Geography*, Blackwell, Maiden, 1998.

Cressey, G. B.: *Asia's Land and Peoples*, McGraw-Hill, New York, 1963.

Critchfield, H. J.: *General Climatology*, Prentice Hall of India Private, New Delhi, 1979.

Dasseman, R. F.: *Environmental Conservation*, John Wiley, New York, 1976.

Demko, G. J.: *Population Geography: A Reader*, McGraw-Hill, New York.

Diem, A.: *Western Europe: A Geographical Analysis*, John Wiley and Sons, New York, 1997.

Dikshit, K. R.: *India: Geomorphological Diversity*, Rawat Publications, Jaipur, 1994.

Doshi, J. K.: *Bhils: Between Societal Self Awareness and Cultural Synthesis*, Sterling, New Delhi, 1971.

Dovering, F.: *Land and Labour in Europe in the Twentieth Century*, L Heinmann, London, 1965.

Doxiadis, C. A.: *Ekistics, An Introduction to the Science of Human Settlements,* Oxford University Press, New York, 1968.

Duckham, A. N. and G. B. Mansfield: *Farming Systems of the World,* Praeger, New York, 1970.

Enyedi, Gyorgy: *Geographical Types of Agriculture,* Budapest, Applied Geography in Hungary, 1964.

Febure, L.: *Geographical Introduction to History,* MacMillan, London, 1925.

Fisher, W. B.: *The Middle East: A Physical, Social and Regional Geography,* Methuen, New York, 1978.

Frankel, F. R.: *India's Green Revolution: Political Costs of Economic Growth,* Princeton University Press, Princeton, 1971.

Ganguli, B. N.: *Trends of Agriculture and Population in the Ganges Valley: A Study in Agricultural Economics,* Methuen, London, 1938.

Gilbert, Alan: *Latin America,* Routledge, New York, 1990.

Gopalakrishnan, Ramamoorthy: *Geography of India,* Jawahar Publishers, New Delhi, 1996.

Gottman, J.: *A Geography of Europe,* Holt, Reinehart and Winston, New York, 1969.

Goudie, A.: *The Human Impact on the Natural Environment,* Basil Blackwell, Oxford, 1986.

Gregor, Howard F.: *Geography of Agriculture: Themes in Research,* Prentice Hall, New Jersey, 1970.

Grove, A. T.: *The Changing Geography of Africa,* Oxford University Press, New York, 1994.

Gugler, J.: *The Urban Transformation of the Developing World,* Oxford University Press, New York, 1996.

Haddon, A.C.: *The Races of Man and their Distribution,* Oxford, New York, 1925.

Hagerstrand, T.: *Innovation Diffusion as a Spatial Process,* University of Chicago Press, Chicago, 1968.

Haggett, P.: *Geography:A Modern Synthesis,* Harper and Row, London, 1972.

Hall, D.: *A History of South East Asia,* St. Martin, New York, 1968.

Haq, M. U.: *Human Development in Changing World,* Oxford, New York, 1992,

Hartshorn, T. A.: *Interpreting the City: An Urban Geography,* John Wiley and Sons, New York, 1990.

Heffman, M.: *Twentieth Century Europe: A Political Geography,* John Wiley and Sons, New York, 1997.

Herbert, D. T. and Colin J. Thomas: *Urban Geography: A First Approach,* Wiley and Sons, New York, 1982.

Hiro, Dilip: *Dictionary of the Middle East,* St. Martin's Press, New York, 1996.

Husain, M.: *Agricultural Geography,* Inter India Publications, New Delhi, 1979.

————————: *Fundamentals of Physical Geography,* Rawat Publications, Jaipur, 2002.

————————: *Geography of Jammu and Kashmir,* Rawat Publications, Jaipur, 2002.

Isard, W.: *Location and Space Economy,* MIT Press, Massachusetts, 1965.

Isard, W.: *Methods of Regional Analysis-An Introduction to Regional Science,* MIT Press, Massachusetts, 1960.

Ismail, Faruqi: *Historical Atlas of the Religions of the World,* MacMillan, New York, 1974.

James, P. E.: *All Possible Worlds: A History of Geographical Ideas,* John Wiley and Sons, New York, 1993.

Jeans, D. N.: *Australia: A Geography,* Sydney University Press, Sydney, 1987.

Jenkins, M. M.: *The Himalayan Kingdoms: Bhutan, Sikkim and Nepal,* Van Mostrand Company, New Jersey, 1963.

Jhingran, A. G.: *Geology of the Himalayas,* Oxford University Press, New Delhi, 1981.

John, M. S.: *The Land and People of China*, Lippincott, Philadelphia, 1989.

Johnson, B.: *India: Resources and Development*, Arnold -Heinemann, London, 1980.

Johnston, R. J.: *City and Society: An Outline for Urban Geography*, Hutchinson, London, 1984.

Jone, H.R.: *A Population Geography*, Harper and Row, London, 1981.

Jones, E.: *Human Geography*, Chatto and Windus, London, 1972.

King, Russel: *The Mediterranean Environment: Environment and Society*, John Wiley, New York, 1990.

Knox, Paul: *Urban Social Geography*, Longman, New York, 1987.

Kothari C. R.: *Research Methodology, Methods and Techniques*, Wiley Eastern, New Delhi, 1990.

Learmonth, A.: *India and Pakistan: A General and Regional Geography*, Methuen, London, 1971.

Lebon, J. H.: *An Introduction to Physical Geography*, Hutchinson, London, 1969.

Martin, G. J. and Geoffrey J.: *All Possible Worlds: A History of Geographical Ideas*, John Wiley, New York, 1993.

Mason, D.: *Theories of Races and Ethnic Relations*, University Press, Cambridge, 1986.

Mathur, S. M.: *Physical Geology of India*, National Book Trust, New Delhi, 1994.

Mehta, J. K.: *Present State of Indian Economy*, Oriental Publishers, Allahabad, 1977.

Minshull, Roger: *An Introduction to Models in Geography*, Longman, London, 1975.

Mukherji, Shekhar: *Dynamics of Population and Family Welfare*, Himalaya Publishing Co., Mumbai, 1991.

Muller, P. O.: *Physical Geography*, Wiley, New York, 1981.

Myrdal, Gunnar: *Rich Lands and Poor*, Harper and Brothers, New York, 1957.

Ojha, S. S.: *Geography of India*, Bhaugolik Adhyayan Sansthan, Allahabad, 2001.

Pannell, C. W.: *East Asia*, Rowman and Littlefield, Lanham, 1999.

Parakh, B. S.: *India: Economic Geography*, N. C. E. R. T., New Delhi, 1990.

Paul, Ward: *World Regional Geography: A Question of Place*, Harper and Row, New York, 1977.

Peach, C.: *Geography and Ethnic Pluralism*, Allen and Unwin, London, 1984.

Persson, G. A.: *Resources and World Development*, John Wiley, Chichester, 1987.

Pohekar, G. S.: *Studies in Green Revolution*, United Asia Publications, Bombay, 1970.

Prescott, J. R.: *Political Frontiers and Boundaries*, Allen and Unwin, Winchester, 1987.

Rampino, M. R.: *Climate: History, Periodicity and Predictability*, Von Nostrand Reinholt, New York, 1987.

Raychaudhuri, S. P.: *Land and Soil*, National Book Trust of India, New Delhi, 1966.

Robert, W. K. and Gilbert, F. W.: *The Environment as Hazard*, Oxford University Press, New York, 1978.

Robillard, A. B.: *Social Change in the Pacific Islands*, Kegan Paul International, New York, 1991.

Rotberg, R.: *Hunger and History: The Impact of Changing Food Production and Consumption Patterns on Society*, Cambridge University Press, New York, 1984.

Rubenstein, J. M.: *The Cultural Landscape*, Prentice Hall, New Delhi, 1990.

Satpathi , D. P.: *Erosion Surfaces, in Outline of Indian Geomorphology: A Study in Regional Geomorphology of Singhbhum*, Classic Publishing, New Delhi, 1981.

Sauer, C. O.: *Agriculture Origins and Dispersals*, American Geographical Society, New York, 1952.

Sealey, Neil: *Caribbean World: A Complete Geography,* Cambridge University Press, New York, 1992.

Sharma R. C: *Readings in General Geography and Geography of India,* Jawahar Publishers, New Delhi, 1992.

Short, J. R.: *Introduction to Urban Geography,* Routledge and Kegan Paul, London, 1984.

Smart, Ninian: *The World's Religions,* Cambridge University Press, New York, 1998.

Smith, C. J.: *China: People and Places in the Land of China,* Sharpe, New York, 1984.

Sopher, D. E.: *The Geography of Religions,* Prentice Hall, Englewood Cliffs, 1967.

Spate, O. H.: *The Spanish Lake: A History of the Pacific Since Magellen,* Beckenham U.K., Croom Helm, 1979.

Srinivasan. T. N.: *The Green Revolution,* Indian Statistical Institute, Calcutta, 1971.

Strahler, A. H.: *Grography and Man's Environment,* John Wiley, New York, 1976.

Sukhwal, B. L.: *Modern Political Geography of India,* Sterling Publishers, New Delhi, 1985.

Taylor, G.: *Geography in the 20th Century,* Methuen, London, 1967.

Taylor, James A.: *Weather and Agriculture,* Pergamon., Oxford, 1967.

Theakstone, W. H.: *The Analysis of Geographical Data,* Heinemann, London, 1970.

Thomas, R.: *Man's Impact on the Environment,* McGraw-Hill, New York, 1988.

Thomson, R. D.: *Processes of Physical Geography,* Longman, London, 1986.

Trewartha, G. T.: *An Introduction to Climate,* McGraw-Hill Book Company, New York, 1954.

Trewartha, Glenn: *The Less Developed Realm: A Geography of its Population,* John Wiley and Sons, New York, 1972.

Turk, J.: *Introduction to Environmental Studies,* C.B.S. College Publishers, Chicago, 1985.

Van Eckelen, W. F.: *Indian Foreign Policy and the Border Dispute,* The Hague, London, 1967.

Wadia, D. N.: *Geology of India,* Tata McGraw-Hill Publishing Co., New Delhi, 1975.

Wadia, F. K.: *Agro-Climatic Regional Planning in India,* Concept Publishing Company, New Delhi, 1996.

Wadia, Mehar D. N.: *Minerals of India,* National Book Trust, New Delhi, 1994

Wegener, A.: *The Origin of Continents and Oceans,* John Biram, New York, 1966.

Willens, A. and Adams, R.: *Dry Lands, Man and Plants,* Architectural Press, London, 1978.

William, H.: *Contemporary Europe: A Geographic Analysis,* John Wiley and Sons, New York, 1997.

Woolley, Leonard: *Prehistory and the Beginnings of Civilization,* Allen and Unwin, London, 1963.

Wurfel, D.: *South East Asia in the New World Order,* St. Martin's Press, New York, 1990.

Zelinsky, W.: *A Prologue to Physical Geography,* Prentice Hall, Englewood Cliffs, 1966.

——————: *A Prologue to Population Geography,* Prentice Hall, New Jersey, 1966.

——————: *Geography and a Crowding World,* Oxford University Press, New York, 1970.

Zeuner, F.E.: *A History of Domesticated Animals,* Hutchinson, London, 1963.

Zhao, S.: *Physical Geography of China,* John Wiley and Sons, New York, 1986.

Ziegler, E. N.: *Encyclopedia of Environmental Science and Engineering,* Gordon and Breach Science Publishers, New York, 1976.

Zohry, D. and M. Hopf: *Domestication of Plants in the Old World,* Clarendon Press, Oxford, 1988.

Index

H

I

N

O

P

T

U

V

W

□□□